The Homeowner's Guide to BUILDING WITH CONCRETE, BRICK & STONE

The Homeowner's Guide to BUILDING WITH CONCRETE, BRICK & STONE

The Portland Cement Association

Rodale Press, Emmaus, Pennsylvania

Printed in the United States of America on recycled paper containing a high percentage of de-inked fiber.

Senior Editor: Margaret Lydic Balitas
Editors: J. William Jones, Larry G. McClung
Editorial Assistants: Paula Bakule, Valerie Patterson
Technical Reviewers (Portland Cement Association): Bruce D. McIntosh, William C. Panarese, Steven H. Kosmatka
Photography: Portland Cement Association except for the following: photos on pages 9, 10, 11, 13, 101, 103, and 104, courtesy of the Rodale Press Photography Department

Tools for photographs on pages 9 and 13 courtesy of Beth-Hanover Supply Co., Bethlehem, Pennsylvania

Book Designer: Darlene Schneck
Illustrator: John Carlance

Library of Congress Cataloging-in-Publication Data

The Homeowner's guide to building with concrete, brick, and stone.

Bibliography: p.
Includes index.
1. Masonry—Amateurs' manuals. 2. Concrete construction—Amateurs' manuals. I. Portland Cement Association.
TH5313.H66   1988   693'.1   88–18413
ISBN 0–87857–795–5   hardcover
ISBN 0–87857–796–3   paperback

2 4 6 8 10 9 7 5 3 1   hardcover
2 4 6 8 10 9 7 5 3 1   paperback

Contents

Introduction

Concrete and masonry have been around for centuries. From ancient times through the present, building with concrete and masonry has always been a chosen method of construction because of its permanence and durability. And its popularity continues to increase because builders realize that this type of construction yields the best return for any investment of time and money. Along with its durability, concrete and masonry construction has reached an all-time high as an aesthetic architectural material. Advanced technology has led to high-quality materials, labor-saving techniques, and a multitude of products unequaled in the history of building. For these reasons, concrete and masonry are chosen by professional architects and builders for a full range of projects from residential buildings and sidewalks to skyscrapers and interstate highways.

But concrete and masonry construction is not just limited to the professional builder. The basic techniques and practices used in concrete and masonry construction have changed little over the past century. As a homeowner, you can cast slabs, pour steps, build walls, and pave driveways and get the same pleasing results a professional builder would. In order to do this, though, you need to know some of the things the builder knows about the materials and techniques—things that a builder learns through much experience.

The idea behind this book is to give you that builder experience. And although there is no substitute for actual on-the-job experience, you can learn enough through reading and your own experience to get the job done well. This book provides you with the basics of concrete and masonry—it's up to you to practice this easy-to-learn skill and gain the experience you need to tackle most any project.

This *Homeowner's Guide to Building with Concrete, Brick, and Stone* is arranged according to the general headings indicated in its title. The first part of the book focuses on concrete—its properties and characteristics, project design, construction techniques, maintenance and repair, and some small projects to get you started. You'll find information on how to prepare a site for slab construction, how to do formwork, how to mix your own concrete or buy ready-mix. In addition, there is a chapter on finishing concrete, which includes special techniques on how to obtain some of the many finishes possible. A chapter on maintenance and repair of concrete will help you get years of quality service out of your concrete projects.

The latter part of the book focuses on masonry—the art of building with units of substantial materials such as brick, stone, and concrete block. Here, you'll be introduced to the wonderful variety of such material available today. You will also learn how mortar is made and which mortar to use for the type of project you're building. In addition to the materials, there is information on how to design sound masonry structures both with and without reinforcement, how to do mortar work, and even how to apply finishes to the final structure using stucco and plaster.

For those who prefer to begin with the simplest form of masonry, we have provided brick, stone, and concrete block projects that require no mortar. Dry construction can yield beautiful results with little effort and skill, and mistakes made are easy to correct.

As a do-it-yourselfer, you know the value of work well done. You can also appreciate the fact that doing your own concrete and masonry work can save you a lot of money, because most such work is quite labor intensive. But, bear in mind that mastering the most highly skilled aspects of concrete and masonry construction requires a lot of time, patience, and practice.

We trust this book will get you off to a good start because it is based on the cumulative knowledge of professional masons. But, reading this book alone is not enough. You've also got to patiently practice the arts described. One way to speed up the learning process is to watch an experienced mason at work. Better yet, spend some time working alongside an experienced mason, even if you have to pay him for the privilege.

After reading this book, you may decide that some projects—pouring the foundation for a house, perhaps, or putting in a driveway—are simply too big and complicated for you to tackle alone. If so, the time you spent learning about concrete and masonry can still be quite valuable. Understanding the fundamental properties of concrete and the basic practices involved in concrete and masonry work will help you in your dealings with contractors. You will be in a better position to explain to them the work you want done and to properly evaluate the quality of their work.

When selecting a contractor, be sure to ask for references and a list of previous jobs accomplished. If possible, ask to see jobs completed several years before, because some improper concrete practices aren't apparent until after two or three years have passed. Also, to save money, consider doing some of the less-skilled jobs, such as site preparation and cleanup, yourself. If you have the time, there is no point in paying skilled craftsman's wages for an unskilled laborer's work.

All projects for the home require advanced planning and this is true with concrete and masonry work. The best time to plan for concrete and masonry projects is in the winter and spring so you can get construction underway when weather conditions are favorable. And because concrete gains strength over a long period of time and only when temperatures are above freezing, it is wise to build projects a month or so before cold weather sets in. If this isn't possible, there are sections in this book dealing with cold-weather construction techniques.

A final note: in planning your concrete and masonry projects, you must check with the local authorities about permits, regulations, and specifications in your area. They will want to see plans and details about the work you are going to do and they can help you with questions about soil conditions and design specifications.

CHAPTER 1

The Time-Honored Art of Masonry

The origins of concrete and masonry lie in the distant past. Enterprising people in ancient cultures experimented with a variety of cementing agents that would help bind brick and stone together. The Assyrians and Babylonians used clay, the ancient Egyptians lime and gypsum. Probably the most durable of ancient cements were the slaked lime and volcanic ash mixtures developed by the Romans during the time of the caesars.

Despite this venerable history, modern masonry construction is indebted principally to events of the more recent past. The cements and mortars universally in use today can be traced directly back to the work of a British mason during the early 19th century. As part of a search for a more durable mortar to use in lighthouse construction, experiments were conducted that led to the discovery and patenting of portland cement in the year 1824.

Portland cement was first manufactured in the United States in 1871. Since that time the United States has led the way in the engineering, manufacturing, and use of concrete and masonry in the building and construction fields.

The Materials

Before proceding further with our discussion of concrete and masonry, some basic terms need definition. First of all, *concrete* can be defined simply as a mixture of aggregates (sand, gravel, or crushed stone) held together by *cement*. In general, *masonry* refers to any type of construction involving the laying of substantial units—bricks, stones, concrete blocks, or tile—with or without a cementing agent to hold them together. When a cementing agent is employed in masonry, it is referred to as *mortar*. Technically speaking, mortar may be described as a type of concrete. What distinguishes it from other types of concrete is its lack of coarse aggregate. Because it is used in thin layers between larger building units, it employs sand alone as its aggregate.

1

Portland Cement

The type of cement used in modern concrete and mortar mixtures was invented and patented by Joseph Aspdin, a British mason, in 1824. Aspdin called his product "portland" cement because the concrete produced from the cement was the color of the natural limestone quarried on the Isle of Portland, a peninsula in the English Channel. The name has endured and continues to be used throughout the world, though many manufacturers add their own trade or brand names. The first portland cement made in the United States was produced at a plant in Coplay, Pennsylvania, in 1871. (**NOTE:** the terms *cement* and *portland cement* will be used interchangeably throughout the remainder of this book.)

Cement consists of appropriate proportions of silica, lime, iron, and alumina. These components come from raw materials that are crushed, milled, proportioned, and blended. After blending, these materials are fed into the upper end of a kiln where they are fired at temperatures ranging from 2,600° to 3,000°F. This extreme heat changes the raw materials chemically into cement *clinker*. When the clinker is cooled, it is pulverized and small amounts of gypsum are added. The resulting cement powder is *hydraulic;* that is, it reacts chemically with water. As the result of this process, called *hydration,* cement and water combine, set, and harden into a stonelike mass. The function of the gypsum added to the dry cement mixture is to regulate the setting time.

There is no typical portland cement manufacturing plant, but all cement manufacturing processes are basically the same (see illustration 1-1).

Concrete

Concrete is basically a mixture of two components: *aggregates* and *paste.* The paste, comprised of portland cement and water, binds the aggregates (sand and gravel or crushed stone) into a rocklike mass as the paste hardens. Aggregates are generally divided into two groups: fine and coarse. Fine aggregates consist of sand with particle sizes up to $3/16$ inch; coarse

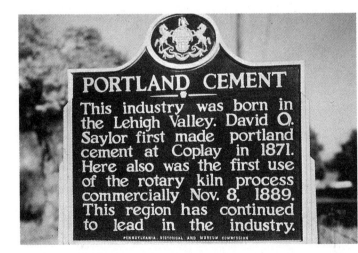

The first portland cement manufactured in the United States was produced in Coplay, Pennsylvania.

This cross section of hardened concrete shows that cement-and-water paste completely coats each aggregate particle and fills all of the spaces between particles.

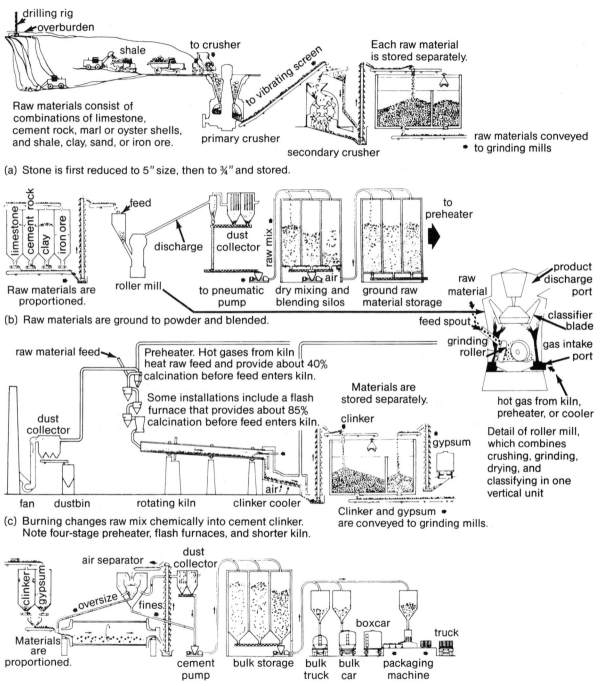

(a) Stone is first reduced to 5″ size, then to ¾″ and stored.

Raw materials consist of combinations of limestone, cement rock, marl or oyster shells, and shale, clay, sand, or iron ore.

(b) Raw materials are ground to powder and blended.

(c) Burning changes raw mix chemically into cement clinker. Note four-stage preheater, flash furnaces, and shorter kiln.

Preheater. Hot gases from kiln heat raw feed and provide about 40% calcination before feed enters kiln.

Some installations include a flash furnace that provides about 85% calcination before feed enters kiln.

Detail of roller mill, which combines crushing, grinding, drying, and classifying in one vertical unit

Clinker and gypsum are conveyed to grinding mills.

(d) Clinker with gypsum is ground into portland cement and shipped.

Illustration 1-1. Steps in the manufacture of portland cement using the dry process.

aggregates are those with particles ranging in size from about $3/16$ to $3/4$ or 1 inch.

Though all cement is basically the same, different cements are manufactured to meet different physical and chemical requirements for specific applications. Differences between cements are the result of variations in the type and quantity of raw material used in the manufacturing process. The American Society for Testing and Materials (ASTM) currently identifies eight types of portland cement (see table 1-1).

Concrete mixtures also can be divided into different classes. As you might suspect, one important factor that distinguishes a particular mixture is the type of portland cement used. But, there's more to it than that. Concrete mixtures also differ with respect to the proportions of cement, water, and aggregates used to create them. Even further differences can be created by the inclusion of various *admixtures*—such as *air-entraining agents* or chemicals that help concrete set up properly in extremely cold weather. Such admixtures will be described in more detail later in this book. (For an example of different types of concrete, see illustration 1-2.)

Table 1-1. **Types of Portland Cement**	
Type I	A general-purpose portland cement suitable for all uses where the special properties of other types are not required. Typical uses include pavements, sidewalks, buildings, bridges, and concrete blocks.
Type II	A specific type of cement used for structures in water or soil containing moderate amounts of sulfate. Also used to moderately control heat buildup in large piers, heavy abutments, and heavy retaining walls.
Type III	A high-early-strength cement with a short curing time. Used when forms need to be removed as soon as possible or when a structure must be put into service quickly. Type III can be used in cold weather because it permits a reduction in the controlled curing time.
Type IV	A special cement used for constructing dams and other massive concrete structures. Generally not available.
Type V	A cement designed to resist chemical attack by soil and water high in sulfates.
Types IA, IIA, and IIIA	These cements are used to make air-entrained concrete. They have the same properties as Types I, II, and III, described above, except that they have small quantities of air-entraining materials combined with them.
White portland cement	A portland cement made from raw materials containing little or no iron or manganese (the substances that give cement its gray color). White portland cement is used primarily in stucco, terrazzo, cement paint, finish-coat plaster, tile grout, and decorative concrete.

Mortars

Centuries ago, combinations of sand and lime were used as mortar. These combinations took months and even years to harden. Consequently, joints between masonry units had to be made quite thin and a lot of painstaking labor was necessary to carefully fit the units together. In the late 19th century, stronger mortars were created by "sweetening" the lime with small amounts of portland cement. The addition of cement caused the mortars to harden more quickly, which allowed thicker joints and more rapid placement of masonry units. This was a great step forward for masonry construction.

The process of adding cement to mortar mixtures eventually led to the development of *masonry cement*—a factory-prepared combination of materials that produce the properties desired in a mortar. Masonry cement includes the following ingredients: portland cement or blended hydraulic cement; a plasticizing material such as finely ground limestone, hydrated lime, or certain clays or shales; air-entraining agents; and sometimes, water-repelling agents. White masonry cement and colored masonry cement containing premilled mineral oxide pigments are also available in many areas. More than 80 percent of all mortar used today is made with masonry cement.

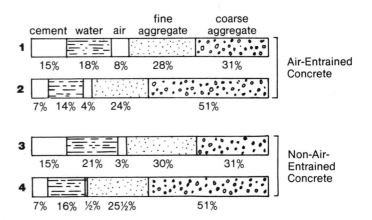

Illustration 1-2. Range in proportions of materials used in concrete, by absolute volume. Bars 1 and 3 represent rich mixes with small aggregate. Bars 2 and 4 represent lean mixes with large aggregate.

Portland Cement Plaster (Stucco)

Portland cement plaster is a combination of portland cement and fine aggregate mixed with water to form a plastic mass able to adhere to a surface and harden, preserving any form and texture it had before setting. Portland cement plaster and *portland cement stucco* are the same material. Typically, the term stucco describes cement plaster used for exterior surfaces; in some areas, however, it refers only to a factory-prepared finish-coat mixture. For ease of understanding, we use the term plaster to mean plaster or stucco.

Bases Portland cement plaster is usually applied in three coats over metal reinforcement with or without solid backing, but two-coat work may be used over solid masonry or concrete. The successful application of plaster to a base depends on the compatibility of the plaster and the base material, the soundness of the base, and the application procedure. Plaster is very compatible with concrete masonry and new concrete, and may be compatible with old concrete or masonry depending on the degree of contamination. When plaster is not compatible with a base, the quality of the base must be upgraded to gain both chemical and

mechanical bond, or metal reinforcement must be furred over the surface. (See illustration 1-3 for methods of applying plaster to different bases. A fairly detailed description of the plastering process is found in Chapter 12.)

Precast Concrete

Precast concrete units, in a wide range of sizes, styles, colors, and textures, are readily available from local precast or concrete masonry producers, home centers, building materials suppliers, or nursery centers. They can be used for building walks, drives, pool decks, patios, and steps. The units are easy to handle and install, requiring only a few tools and little special knowledge of concrete.

Paving Slabs Precast concrete paving slabs and interlocking pavers are machine made or precast by conventional methods. Slabs are commonly available in square and rectangular shapes measuring from 12 to 36 inches, with a thickness of 2, 2½, or 3 inches. The 2-inch thickness is suitable for walks and patios, while the 2½- or 3-inch thickness should be used for

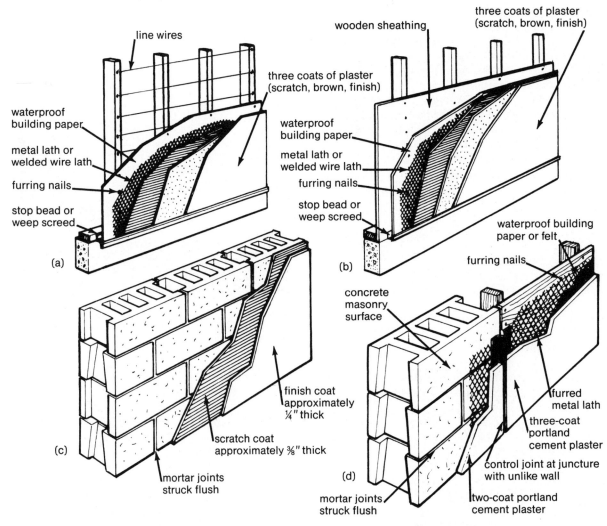

Illustration 1-3. Stucco may be applied on a variety of bases: (a) open wooden framing, (b) sheathed wooden framing, (c) concrete masonry, and (d) unlike bases such as a combination of wood and masonry.

driveways. Besides rectangular slabs, some of the more popular shapes and styles are round, triangular, diamond, hexagonal, and Spanish tile.

Interlocking concrete pavers are growing in popularity and use at a very rapid rate all over the world. They are made in a number of shapes and colors and are ideal for do-it-yourself applications. Concrete pavers interlock without the use of mortar. This allows them to be "unzipped" for work on or beneath the subsurface, then "zipped" back using the same material. Typical paver thicknesses are $2^3/_8$ inches (60 mm) and $3^1/_8$ inches (80 mm). The thinner units are for pedestrian areas and residential driveways, the $3^1/_8$-inch pavers are for streets and other vehicular applications.

Precast concrete rounds can be obtained in various diameters, thicknesses, and surface finishes for use in garden walkways or stepped approaches to your home.

Concrete pavers make very attractive walkways and driveways that can be accentuated with solid concrete borders.

Precast steps are cast all in one piece in a factory. With special equipment, precast steps are lifted from the truck into position in a short time.

Slabs may be a natural gray or white cement color, or pigmented in tones of red, black, brown, green, or yellow. Slabs with exposed-aggregate finishes are available in some areas.

Steps Complete or partial use of precast concrete units can simplify step construction. Precast steps with treads, risers, sidewalls, and porch all cast as one piece, are available in a wide variety of styles and sizes. Widths of 3 to 6 feet with any number of risers up to six, and porches up to 6-feet deep are commonly available.

Precast steps are used for new construction and as step replacements for older homes. The units are delivered by the manufacturer ready for installation. Often the manufacturer can install the units quickly with special trucks and hoists. Railings are bolted to inserts cast into the steps during manufacturing.

Concrete Block

Concrete masonry units (concrete *block* and concrete *brick*) are made mainly of portland cement, graded aggregates, and water. Mass production has contributed to the relatively low cost of concrete masonry units, and in many production plants, some phases of the manufacturing process are completely automated.

Concrete masonry units are available in sizes, shapes, colors, textures,

This house is an example of the tremendous architectural potential of concrete masonry units. Besides such functional uses as foundations and footings, concrete block are used extensively as an integral feature in buildings and landscaping.

and profiles for practically every conceivable need and convenience in masonry construction. (See Chapter 10 for a detailed description of different types of block.) Because of this variety, concrete masonry units may be used to create attractive patterns and designs to produce an almost unlimited range in architectural treatments of wall surfaces. The list of applications is lengthy, but some of the more prominent uses are for the following:

- Backing for brick, stone, stucco, and other exterior facings
- Bond beams, lintels, and sills
- Chimneys and fireplaces
- Piers, pilasters, and columns
- Retaining walls, slope protection, and ornamental garden walls
- Walls (exterior, interior, load bearing, and nonload bearing)
- Pavements and steps

Brick

Bricks are made of clay that has been fired in a kiln to produce a dense and durable substance. Bricks have been around for thousands of years and as a result, there are literally thousands of different sizes, shapes, textures, and colors. This wide variety breaks down into four basic types, the first of which, building brick, accounts for nearly 98 percent of all brick used.

Building Brick Building bricks are red and rectangular in shape. They are produced in great quantity and are therefore the least expensive of all brick. Within this general type, there are three grades: nonweathering (NW), moderate weathering (MW), and severe weathering (SW). Nonweathering brick is intended for interior use only, whereas moderate-weathering brick can be used outdoors in southern climates and severe-weathering brick can be

Brick is available in a wide variety of shapes, types, and sizes. Here are four examples: (clockwise, left to right) a cored building brick, a firebrick, a decorative face brick, and a paving brick with chamfered edges.

Paving brick can be arranged in many different patterns to create yard and garden pathways that are both attractive and functional.

used in any environment. The differences between these grades of brick are the result of the temperature at which they are fired. Firing bricks fuses the clay, causing it to vitrify. The greater the vitrification, the less porous the brick. When bricks are porous, they absorb water and this water causes the brick to break apart when temperatures drop below freezing.

Face Brick Face bricks are made for decorative purposes and come in a wide variety of texture, color, and dimension. For the most part, they are rated SW and are made with more care than building brick, which also makes them more expensive. In fact, they are the most expensive type of brick made. As their name implies, they are only used for exterior surfaces.

Firebrick Firebricks are used to line fireplaces and furnaces or any location where protection from heat is required. Made from special clays and fired to higher temperatures than building or face brick, firebrick can withstand intense heat. Building with firebrick requires the use of fireclay mortar, be-

Brick is also a good choice for creating low garden walls suitable for sitting and enjoying the scenery.

Brick pavers set on beds of gravel and sand and surrounded by treated lumber make beautiful landings along stepped pathways up steep embankments.

This brick privacy wall surrounding the backyard of a contemporary suburban home features the weeping mortar look described in Chapter 7, illustration 7-2.

cause conventional mortar disintegrates when exposed to intense heat.

Paving Brick   Paving bricks are harder and more durable than building bricks and are sized for use without mortar. Paving bricks are typically used for driveways, parking areas, and sidewalks, where they are laid on a bed of sand.

Stone

Stone is the oldest of all building materials. The most common types of stone used in masonry are limestone, sandstone, granite, and slate. Stone is denser and heavier than most other building materials, and because its use requires a great deal of preparation, it is mostly limited to smaller projects such as low fences, walkways, wall veneers, or hearths. Stone is referred to by more specific names based on its characteristics: *ashlar* is cut stone; *rubblestone* (or rubble) is uncut stone; *flagstones* are flat stones used for paving; *cobblestones* are naturally rounded stones; and *fieldstones* are rocks found on the ground.

Limestone and Sandstone Limestone and sandstone are *sedimentary* rocks. Sedimentary rocks are formed in layers from sediment, such as sand and seashells, over a period of millions of years. Because these rocks were formed in layers, they split easily along the sediment lines. Both limestone and sandstone are used extensively for paving, fences, or as building veneer.

Granite Granite is an *igneous* rock formed deep in the earth from molten magma during the early stages of the earth's development. Igneous rocks are heavy, extremely hard, and quite durable. Rough-cut granite is used for paving or as building veneer.

Marble and Slate Marble and slate are *metamorphic* rocks, a compact and highly crystalline type of rock that results when rocks of various kinds are subjected to the combined forces of water, extreme heat, and pressure. The earth's heat and pressure are so great that identifying the original type of rock is often impossible. Slate is commonly used for floors, walks, or roofs. Polished marble is used for floors, fireplaces, or commercial applications.

Ordering Materials

Before beginning any concrete or masonry project around your home,

Fieldstone is perennially popular as a material for building retaining walls, such as the one shown here.

Interesting designs can be created by combining different types of masonry units. Here we see a brick pathway and steps surrounded by mortarless stone walls.

Flagstone pathways can be created without mortar or with mortar, as in this example.

check on the availability of materials in your area. The range of concrete masonry units, precast products, brick, and stone varies from one part of the country to another and from one distributor to another. Start your search by checking the yellow pages of your phone book, then visit home centers and local building suppliers to see what products they handle as well as the types of products they can order. If you pass a construction site where a particular material is being used that interests you, ask the foreman where the material came from.

Proper purchasing of materials requires advanced planning. Whether you buy large or small quantities, it helps to know what your buying options are and how to order. While working on your design, find the most economical way to purchase supplies: you can get price breaks on brick and block, ready-mix concrete, and stone when you buy in certain lot sizes. When finalizing your design, take into account the most economical quantities of the various materials you intend to use.

In drawing up your order, also keep in mind that you need to add an extra 5 to 10 percent to the final quantity calculations in your design to allow for breakage and waste. Ordering extra materials the first time around not only is more economical than making two orders, it also provides insurance against time-consuming delays in the completion of a project. Moreover, it improves the chances that all materials used in the project will be from the same production *batch*, and thus have consistent qualities and properties. A final point to remember when pricing and ordering materials: don't forget to inquire about the cost of delivery and any other incidental charges.

Tools

Masonry, like most arts, involves more than an intelligent mind and an experienced hand. It also involves the use of a wide range of tools. Some of them—such as the pointing *trowel,* the V-jointer, the star drill, and the stone chisel—have evolved specifically to meet the needs of various masonry-related tasks. Others—such as the square shovel, the wheelbarrow, and the level—meet the needs of nonmasons as well as masons. A good set of basic tools, both generic and specialized, are necessary aids to doing quality masonry work.

At this early point in the book, we are not going to go into lengthy verbal descriptions of specific tools and their uses. Instead we are inserting some photos that identify the principal tools mentioned elsewhere in this book. Further discussion of these tools will occur when the tasks they are designed to do are discussed.

Obviously, even the finest and most expensive of tools, if wielded by a clumsy amateur, is not likely to produce quality work. Likewise, a highly skilled craftsman can often do amazing work with less than the best of tools. Thus, filling your home workshop with a complete line of mason's tools will not in itself guarantee you are well on your way to mastering the art of masonry. At the same time, you should not underestimate the importance of tools. Becoming familiar with the tools that are available and learning which ones are best suited to the tasks you wish to accomplish will be a great aid to you as you get started in masonry work. As you begin to think about tools and their acquisition, there are a couple of rules of thumb that are good to observe.

First. No matter which craft you want to learn, it's seldom a good idea

These are the basic tools of the trade for hand finishing concrete. At the top is a straightedge, and just below, a darby. From the left is a finish trowel, a jointer (or groover), an edger, and a magnesium float.

to begin the learning process by rushing out and buying every tool that has been invented for that skill. Start slowly. Don't buy anything until you have a specific project planned and are ready to go. Then buy only those tools you know you will need to complete that project. As time passes and you move from one project to the next one, add more tools to your collection as you need them. This way, you will acquire only those tools that you actually need.

Second. It usually pays to buy the best quality tool you can afford. This is especially true if you know you will use the tool on numerous occasions and are prepared to take care of it so it will last a long time. However, if you encounter a task that you know you will never attempt a second time and an inexpensive tool will prove ade-

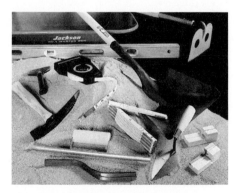

Here are some common tools used in masonry work: mortar hoe, mortar pan, square shovel, level, 50-foot tape, modular rule, brick set, bricklayer's hammer, pointing trowel, mason's line blocks, and a pair of joint finishing tools.

quate to complete the job, then you are probably better off going with the inexpensive tool.

Illustration 1-4. Protect your eyes at all times when working with concrete. Protect your head any time you work under ladders and scaffolding. Protect your skin by wearing protective clothing, rubber boots, and waterproof gloves. Remove contaminated clothing as soon as possible.

When you do buy a tool that is worth using more than once, by all means take good care of it. If it comes into contact with concrete or mortar, clean it carefully at the end of each day and put it away in a place where you will be able to find it quickly the next time you want to use it. After cleaning, tools that are prone to rust should be coated with a rust-preventing oil and stored in a dry place. If you have ever experienced the disappointment of reaching for a tool that has been idle for many months or years, only to discover it has rusted beyond usefulness, you can appreciate the importance of such preventive maintenance. Good tools are made to last a lifetime and will if they are cared for properly.

Safety

Concrete is one of the safest building materials known. Over the years, relatively few people involved in mixing, handling, and finishing concrete have experienced injury. In order to continue this tradition of safety, here are a few precautions you should consider while working with portland cement, mortar, and concrete mixtures.

Protect your eyes and head. Eye protection is essential to guard against blowing dust, concrete that may splatter, or other foreign objects. Wear proper eye protection when working with cement or concrete. Depending on the job, eye protection should consist of safety glasses with side shields or goggles (see illustration 1-4).

Head protection, while critical on large building projects, is not necessarily required on most home jobs. But, when working on scaffoldings or ladders with brick, stone, or block, it's important to keep others away from below your work area.

Protect your back. A short-handled, square shovel is the most effective tool for spreading concrete. After the concrete is deposited in the desired area by truck-mixer chute or wheelbarrow, it should be pushed, not lifted, into final position with the shovel.

When moving or lifting heavy bags of materials, lift the bags carefully to avoid putting any strain on your back muscles. If possible, get a helper to assist you in moving bags of cement, because a single bag typically weighs 94 pounds.

Protect your skin. When working with fresh concrete or mortar, care should be taken to avoid skin irritation or chemical burns that can occur through prolonged contact between the fresh mix and skin surfaces, eyes, and clothing. Three factors cause skin irritation or chemical burns through prolonged contact with fresh concrete or mortar.

• Portland cement is alkaline in nature and therefore caustic.

• Portland cement is hygroscopic, meaning it tends to absorb moisture. It will absorb moisture from your skin.

• Sand contained in fresh concrete or mortar is abrasive to bare skin.

Clothing should not be allowed to become saturated with the moisture from the concrete because saturated clothing can transmit alkaline or hygroscopic effects to the skin. To protect your skin, wear waterproof gloves, a long-sleeved shirt, and full-length trousers.

If it is necessary to stand in fresh concrete while it is being placed and screeded, or floated, wear rubber boots high enough to prevent concrete from flowing into them (see illustration 1-4).

When working with fresh concrete or mortar, begin each day by wearing clean clothing, and conclude the job or day with a bath or shower. During the finishing process, waterproof pads should be used between fresh concrete surfaces and knees or elbows. Clothing areas that become saturated from contact with fresh concrete should be rinsed out promptly with clean water to prevent continued contact with skin surfaces. Eyes or skin areas that come in contact with fresh concrete should be washed thoroughly with fresh water. Mild irritation of skin areas can be relieved by applying a lanolin cream to the irritated area after washing. Persistent or severe discomfort should be attended by a physician.

Becoming familiar with the basic tools and materials and understanding how to handle them safely are important first steps toward learning the time-honored art of masonry. From this point on, it's just a matter of undertaking some projects and building up your experience.

CHAPTER 2 # Understanding Concrete

Concrete is basically a mixture of aggregates and paste. This paste, consisting of portland cement and water, binds the aggregates (sand and gravel or crushed stone) into a rocklike mass as the paste hardens due to the chemical reaction of the cement and water (hydration).

The performance of concrete is related to workmanship, mix proportions, aggregate characteristics, amount and type of cement, presence or absence of entrained air, and adequacy of curing. When these factors are not carefully controlled, they may adversely affect concrete's *workability* and eventually its strength and durability.

Qualities of a Concrete Mixture

Strong and durable concrete walks, driveways, patios, and steps are constructed from good-quality materials placed with care and competence in each step of construction. Good-quality concrete should be workable, tolerant of some misuse, strong enough to support applied loads, and durable under the most severe weather conditions. To create a workable, strong, and durable concrete mixture, the proper type of concrete and the proper proportions of ingredients must be chosen to fulfill the requirements of the project.

Workability

Workability is the ease with which freshly mixed concrete can be placed, compacted, and finished with no segregation or separation of the individual components. Workability varies according to the application. For example, concrete with workability suitable for a sidewalk would be difficult or even impossible to place in a thin, heavily reinforced wall.

Plasticity and *consistency* of the concrete are important to workability and have a major influence on the serviceability and appearance of the finished concrete. Consistency affects the ease of placing and molding the concrete and is measured by slump. Concrete mixes with moderate consistency are readily placed by hand methods. Stiffer mixes require mechanical vibration to compact the concrete.

Slump is a measure of the consistency of a concrete mixture—the higher the slump, the wetter the mixture. Slump is indicative of workability when assessing similar mixtures. When used with different batches of the same mix, a change in slump indicates a change in consistency and in the characteristics of materials, mix proportions, or water content.

A significant change in slump from batch to batch indicates that a change has occurred in the mix, and corrections must be made to restore uniformity. Adding a little water to restore slump is permitted, but it is a poor practice because too much water weakens concrete, lowers its resistance to freezing and *deicer* salts, and results in shrinkage cracks as the concrete hardens.

When ordering ready-mix concrete, ask for a slump of 3 to 5 inches, which is suitable for most home projects.

A slump test is made by filling a cone with concrete and using a rod to assure proper compaction. Excess concrete is scraped away and the cone slowly removed. Finally, the cone is placed beside the concrete and the slump is measured using a rod and a ruler.

Strength

The principle factors affecting the strength of concrete are the water-cement ratio, age, and adequacy of *curing* (the extent to which hydration has progressed). Simply put, the more water used in making concrete the weaker concrete becomes, and the longer concrete cures, the stronger it becomes. Because the workability of concrete is directly related to the amount of water in the mix, the right balance must be found between strength and workability. Care must be taken to limit the water content to just that amount necessary to achieve a workable mix. Adding more water than that will produce weaker concrete.

Strength in concrete refers to compressive strength and is measured in terms of pounds per square inch (psi). Compressive strength is measured from a sample of a mix that has been cast into a standard 6-inch-diameter by 12-inch-high cylinder, allowed to cure for 28 days, then broken using a laboratory testing machine. When ordering ready-mix, you may hear the concrete referred to as 4,000 psi-mix or 4,000 psi-concrete. This means a standard cylinder of this concrete can withstand up to 4,000 pounds of pressure per square inch after curing for 28 days. Normal mixing proportions and procedures will yield concrete of adequate strength for most home projects. However, for resistance to freeze—thaw cycles and deicer chemicals, make sure your concrete is a 4,000 psi air-entrained mix.

Durability

Durability is the ability of concrete to resist weathering action, chemical attack, abrasion, and other conditions of service. Among factors affecting durability are the freezing and thawing of water in concrete, the amount of entrained air in the concrete, the selection of aggregates, and the degree of air drying before freezing. Scaling is the most common sign that concrete is not durable. (Factors affecting durability are discussed later in this chapter.)

Ingredients of a Concrete Mixture

The components of concrete—cement, water, sand, and gravel or crushed stone—are all critical to the quality of the final product. Good-quality concrete costs no more to make than poor-quality concrete, but is far more economical in the long run because of its greater durability. The rules for making good-quality concrete are simple.

- Use proper ingredients.
- Proportion ingredients correctly.
- Measure ingredients accurately.
- Mix ingredients thoroughly.

Cement

Portland cement is not a brand of cement but a type. Most portland cement is gray in color; however, white portland cement is manufactured from special raw materials that produce a pure white color. White portland cement can be used instead of the normal gray portland cement; but it costs more. That's why it is generally used only for decorative work and other special jobs.

You can buy portland cement in bags at your local building materials center. In the United States, a *bag* of cement weighs 94 pounds and holds 1 cubic foot. By contrast, a Canadian bag of cement weighs 40 kilograms, or approximately 88 pounds.

Portland cement is a moisture-sensitive material: if it is kept dry, it will retain its quality indefinitely, but if it is stored in contact with moisture,

it will set more slowly and have less strength than cement that is kept dry. The relative humidity in a garage or shed used to store cement should be as low as possible. Cement bags should not be stored on damp floors, but should rest on pallets. Bags should be stacked close together to reduce air circulation and never stacked against outside walls. Bags stored for long periods should be covered with tarpaulins or other waterproof covering.

On smaller jobs where no shed is available, bags should be placed on raised wooden platforms. Waterproof coverings should fit over the pile and extend over the edges of the platform to prevent rain from reaching the cement and the platform. Rain-soaked platforms can damage the bottom bags of cement.

Cement stored for long periods may develop what is called *warehouse pack*. This can usually be corrected by rolling the bags on the floor. Cement suitable for use in concrete should be free flowing. The presence of lumps that cannot readily be pulverized between your thumb and finger indicates that the cement has absorbed moisture. Such cement should never be used for important work; but when the lumps have been screened out through an ordinary window screen, it can be used for certain minor jobs such as setting fence posts.

Different types of portland cement are manufactured to meet different physical and chemical requirements. There are eight basic types of cement (see table 1-1). For most residential projects, select either Types I, IA, II, IIA, III, or IIIA for the best all-around results.

Water

Almost any natural water that is drinkable and has no pronounced taste or odor can be used as mixing water for making concrete. Although some waters that are not suitable for drinking will make satisfactory concrete, to be on the safe side, use only water fit to drink.

Aggregates

Aggregates are minerals such as sand, gravel, and crushed stone that make up 60 to 80 percent of concrete by volume. They act as an inert filler material to reduce the amount of cement required in concrete. Without them, concrete would be very expensive. Furthermore, aggregates restrain the shrinkage that occurs when concrete hardens. This is important, because shrinkage can lead to excessive cracking.

Aggregates are divided into two types: *fine* and *coarse*. Fine aggregate is usually natural sand and coarse aggregate is usually gravel or crushed stone. Aggregates within these two types range in size from small to large in such a way that smaller particles fill in voids between larger sizes.

Natural sand is the most commonly used fine aggregate, although manufactured sand, made by crushing gravel or stone, is also available in some areas. Sand should have particles ranging in size from $1/4$ inch down to dust-size particles. Mortar sand should not be used for making concrete because it contains only small particles.

Gravel and *crushed stone* are the most commonly used coarse aggregates. They should consist of particles that are sound, hard, and durable—not soft or flaky—with a minimum of long, sliverlike pieces. Particles should range in size from $1/4$ inch up to a maximum of $3/8$, $1/2$, $3/4$, 1, or $1^1/2$ inches. The most economical concrete mix is obtained by using coarse aggregate with the largest practical maximum size. In slabs for driveways, sidewalks, and patios, this would be about one-third the slab thickness. Accordingly, 4-inch-thick slabs may use coarse aggregate with 1-inch maximum size, whereas slabs of 5- to 6-inch thickness may use $1^1/2$-inch maximum-size aggregate. Generally, a 1-inch maximum-size aggregate should be used for steps.

Actually, in most areas, ready-mix will contain aggregate with a top size of either $3/4$ or 1 inch. Thus, although you should use the maximum practical size aggregate in your concrete mix, you may not always find it available.

Gravels are more or less smooth and rounded whereas crushed stone aggregates are rougher and more angular. Angular particles produce mixtures that are a little more difficult to work than mixtures with rounded particles, hence a little less crushed material must be used in each cubic foot of concrete to get the same workability.

Selecting Aggregates Both fine and coarse aggregates for making concrete must be clean and free of excessive dirt, clay, silt, coal, or other organic matter such as leaves and roots. These foreign materials will prevent the cement from properly binding the

Well-graded aggregates have particles of various sizes. Shown here is $1^1/2$-inch maximum-size coarse aggregate. Particles vary in size from $1/4$ to $1^1/2$ inches.

aggregate particles together, resulting in porous concrete with low strength and durability.

Good fine and coarse concrete aggregates have a full range of size from the smallest to the largest, but no excess amount of any one size. The big particles fill out the bulk of a concrete mix, and the smaller ones fill in the spaces between the larger ones. Aggregates with an even distribution of particle sizes are said to be well graded. Such aggregates produce the most economical and workable concrete. Do not use mixtures of fine and coarse aggregates taken directly from gravel banks or stone crushers. These aggregates usually contain an excess of sand in proportion to coarse material. Before using this material in concrete, it should be screened and recombined into properly graded fine and coarse aggregates.

When *batching* your own concrete, buy fine and coarse aggregates separately from a reputable building materials center. If there is a ready-mix producer in your area, it is preferable to purchase aggregates there. Such a producer will make sure that the aggregates you buy have the correct sizes and are suitable for making concrete.

Store aggregates on a clean hard surface, if possible, and not directly on the ground where they may be contaminated by dirt or mud. Cover aggregate piles with canvas or plastic tarpaulins to prevent them from becoming wet in case of rain. Do not use the bottom layer of an uncovered aggregate pile because this part is usually saturated with water and may contain an accumulation of dirt washed through from higher layers.

Entrained Air

Air is also an important ingredient for making good concrete. In the late 1930s it was discovered that air, in the form of microscopic bubbles evenly dispersed throughout the concrete, improved durability and virtually eliminated scaling due to freeze-thaw cycles and deicer salt action. Concrete containing such air bubbles is called *air-entrained concrete.*

Unlike entrapped *air voids*, which occur in all concretes, intentionally entrained air bubbles are extremely small in size—less than 4 thousandths of an inch. The bubbles are not interconnected and are well distributed throughout the paste.

Air entrainment is most important for concrete exposed to severe weather. In cold climates, and even in moderate climates that have several freeze-thaw cycles each year, air-entrained concrete should be used for all exterior work, including driveways, sidewalks, patios, and steps.

Here's how air entrainment works. Hardened concrete usually contains some water. When this water freezes, it expands causing pressure that can rupture (scale) the concrete surface. The tiny air bubbles act as reservoirs or relief valves for the expanding water, thus relieving pressure and preventing damage to the concrete.

Besides enhancing concrete's durability, air entrainment has other advantages. For example, the tiny air bubbles act like ball bearings in the mix. They increase its workability and cohesiveness with the result that less mixing water is required.

To create air-entrained concrete, chemicals specifically made for this purpose, called *air-entraining agents,* are added to the mixing water. (Ready-mix producers will sometimes provide small quantities of air-entraining agents to do-it-yourselfers.) Many cement manufacturers also produce portland cements that have air-entraining agents ground together with the cement. Cements of this type have the designation A following the type number. For example, Type IA cement is Type I portland cement that contains an interground air-entraining agent.

When ordering ready-mix concrete, ask for an air content of 5 to 7$\frac{1}{2}$ percent depending on your maximum aggregate size (see table 2-2). When batching your own concrete, follow the manufacturer's instructions concerning the amount of air-entraining agent to add to the mix. Or, to avoid the trouble of buying and measuring an air-entraining agent, ask your materials supplier for a type of portland cement that has an interground air-entraining agent.

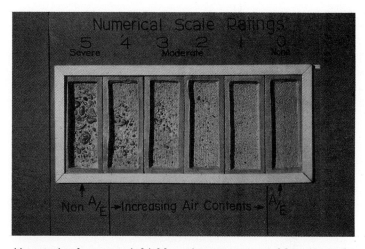

Air-entrained concrete is highly resistant to repeated freezing and thawing cycles. These specimens were subjected to numerous cycles of freezing and thawing with application of deicer chemicals. The specimen at the left is non-air-entrained and has severely scaled. The specimen at the right is air-entrained and has not scaled.

Workability Agents

Fresh concrete is sometimes *harsh* (an industry term meaning hard to work) as a result of certain aggregate characteristics such as particle shape and improper grading or faulty mix proportions. Under these conditions, improved workability may be needed, especially if the concrete requires a troweled finish. Frequently, increasing the cement content or the amount of fine aggregate will give the desired workability.

The best workability agent is entrained air. As previously described, it acts like a lubricant and is especially effective in improving the workability of lean, harsh mixtures. Other workability agents consist of organic materials or finely divided minerals.

Other Admixtures

Concrete should be workable, finishable, strong, durable, watertight, and wear resistant. These qualities are easily and economically obtained by carefully selecting suitable materials. There may be instances, however, when special properties are required. Properties such as extended time of set, acceleration of set, and early strength development may be obtained by using admixtures, ingredients added to concrete either immediately before or during mixing. Air-entraining agents are the most important admixtures and have already been described. Here is a brief description of other admixtures sometimes employed.

• *Water-reducing admixtures* are used to reduce the amount of mixing water needed to produce concrete of a given consistency or to increase the slump of concrete for a given water content.

• *Retarding admixtures* retard the setting time of concrete from 30 to 60 percent and are sometimes used to offset the accelerating effect of hot weather on the setting of concrete.

• *Accelerating admixtures* accelerate strength development at an early age.

We will not go into more detail about these admixtures for a couple of reasons: (1) they are not likely to be needed on projects you will undertake around the home and (2) each of them has drawbacks—such as increasing shrinkage or reducing the strength in concrete—that make their use problematic. If you do decide to use one of these admixtures, make trial mixes under the same conditions of humidity and temperature that will be present when you use the concrete. And, use only the amount recommended by the manufacturer or the optimum amount determined by your trial batch. (If you want more information on admixtures, contact the Portland Cement Association.)

Ready-Mix Concrete

Usually, the most convenient and economical source of concrete is a ready-mix producer, who can supply concrete to meet the requirements of any project. Ready-mix concrete is sold by the cubic yard (27 cubic feet), and

Ready-mix concrete is ideal for most jobs. Accurately proportioned and loaded at the plant, it is mixed in the truck during transport.

a producer will usually deliver any quantity greater than 1 cubic yard. Whether the project requires 1 cubic yard or many, the job of proportioning, weighing, mixing, and hauling will be done according to careful specifications to ensure concrete of uniform quality.

The cost of ready-mix concrete varies according to the distance hauled, size of the order, day of delivery, unloading time, and type of mix. A call to one or more reputable producers should be made to establish the local price. If the quantity of concrete needed is less than 1 cubic yard or if there is no ready-mix plant in the area, it will be necessary to make the concrete at the job site.

Estimating Concrete Requirements

Estimating the amount of concrete you need to order can be done by using table 2-1 and the example that follows.

To find the amount of concrete required for a 4-inch-thick driveway 11 × 41 feet, for example, first figure the number of square feet. Multiplying 11 by 41 you get an area of 451 square feet. This can be rounded off to 450. Now, turn to table 2-1 and note the following:

300 square feet = 3.70 cubic yards
100 square feet = 1.23 cubic yards
50 square feet = 0.62 cubic yard

Add these three amounts together and you learn that for 450 square feet you need 5.55 cubic yards of concrete.

With a perfect subgrade and no losses from spillage, 5½ cubic yards might be enough. But for insurance against contingencies, the order should be increased to 6 cubic yards. It is always better to have some concrete left over than to run short.

How to Order

The following five points should be considered when ordering ready-mix concrete.

- **Maximum-size aggregate** should not exceed one-third the slab thickness. Accordingly, 4-inch slabs can use aggregate no larger than ³/₄ or 1 inch. Slabs of 5 to 6 inches can use aggregate up to 1¹/₂ inches, if it is available. A 1-inch top aggregate size is suggested for steps.

- **Minimum cement content** should be no less than the amount given in table 2-2 for the particular maximum-size aggregate used. These cement contents are essential for proper finishing and strength development. In areas exposed to freeze-thaw cycles or to deicers, it is advisable to use a minimum cement content of 564 pounds (6 bags) per cubic yard.

- **Maximum slump** should not exceed 5 inches. Actually, a 4-inch slump will give a good, workable mix. Stiff mixes are harder to place and finish by hand, but can be used to advantage with mechanical placing and finishing equipment. Very wet, soupy mixes will not make durable concrete.

- **Compressive strength** at 28 days should be no less than 3,500 psi, unless the concrete will be exposed to freeze-thaw cycles and deicers. In that case, 4,000-psi concrete is required. In some instances, cement contents higher than those given in table 2-2 may be necessary to obtain 3,500-psi or 4,000-psi concrete.

- **Air content** given in table 2-2 is required to obtain good durability in all concrete exposed to freezing and thawing and deicing salts. When this protection is not required, air content about one-third less is useful to reduce *bleeding* and segregation and improve workability and finishability.

Table 2-1. Estimating Cubic Yards of Concrete for Slabs

Thickness, in.	Area in Square Feet (width × length)					
	10	25	50	100	200	300
4	0.12	0.31	0.62	1.23	2.47	3.70
5	0.15	0.39	0.77	1.54	3.09	4.63
6	0.19	0.46	0.93	1.85	3.70	5.56

NOTE: Does not allow for losses due to uneven subgrade, spillage, etc. Add 5 to 10 percent for such contingencies.

Table 2-2. Guide for Ordering Ready-Mix Concrete for Drives, Walks, and Patios

Maximum-Size Aggregate, in.	Minimum Cement Content, lb. per cu. yd.	Typical Slump, in.	Compressive Strength at 28 Days, lb. per sq. in.*	Air Content, percent by volume
³/₈	610	3 to 5	3,500	7¹/₂ ± 1
¹/₂	590	3 to 5	3,500	7¹/₂ ± 1
³/₄	540	3 to 5	3,500	6 ± 1
1	520	3 to 5	3,500	6 ± 1
1¹/₂	470	3 to 5	3,500	5 ± 1

NOTE: Slump should neither exceed 5 inches nor fall below 3 inches.
*Use 4,000 psi for freeze-thaw and deicers.

In addition to the above, the ready-mix producer needs to know the number of cubic yards required, as well as where and when to deliver the concrete. If possible, orders should be placed at least a day ahead of time but 2 or 3 days is preferable.

Ready-Mix Ordering Tips Most ready-mix producers already have the mix you want for your home projects. But there are some things you need to consider before ordering. Here's a checklist.

• Know how many cubic yards of concrete you'll need, and be sure this includes 5 to 10 percent extra to allow for contingencies.

• Have helpers and/or cement masons lined up to be on site when the ready-mix truck arrives.

• Listen to weather reports to assure favorable conditions.

• Make sure all *forms* are in place and ready to go.

Then, when you call in your order, provide the dealer with the following information:

• Cubic yards of concrete required

• Type of project—sidewalk, steps, basement floor, garage floor, driveway, foundation, and so on (be sure to specify air-entrained for exterior work)

• Distance to worksite from the closest a truck can park (this permits the ready-mixer to provide the appropriate number of chutes)

Also, when you order, ask for the following information:

• Cost per cubic yard

• Cost for delivery

• Amount of time driver can be on site before additional charges start

• Additional charge for driver's time

• Method of payment (cash, check)

Mixing Your Own Concrete

Proportioning Ingredients

The quality of the concrete depends on the quality of the cement paste. The quality of the paste, in turn, depends on the amount of water mixed with the cement and the extent of curing. If too much water is used, the paste will be thin and diluted, making the concrete weak and porous. As the amount of water is reduced, the strength of the paste increases, making the concrete stronger and more durable.

To find the correct amount of mixing water, use the proportions given in tables 2-3 and 2-4 as a starting point. These proportions may not always yield a workable mix with your aggregates, but they do serve as a starting point.

The weights of material given in table 2-3 will make a 1-cubic-foot batch. This is about the right amount for hand mixing. For machine mixing, multiply the values in the table by the capacity of the mixer. For example, for a 3-cubic-foot mixer, multiply each ingredient by 3.

The proportions given in table 2-4 are by volume (or parts) and can be measured in pails, cans, or any other sturdy container. An ordinary galvanized water pail or a 5-gallon plastic container makes a convenient batching container.

Estimating Quantities Needed

Before getting down to the job of measuring and mixing, you'll need to know just how much cement, sand, and coarse aggregate to buy for your project. To do this, first estimate the amount of concrete your project will require. Use the following simple formula, which works for any square- or rectangular-shaped area.

$$\frac{width\,(ft) \times length\,(ft) \times thickness\,(in)}{12} = cubic\,feet$$

For example, a 3-foot-wide sidewalk that is 20 feet long and 4 inches thick would require 20 cubic feet of concrete ($[3 \times 20 \times 4] \div 12 = 20$). This doesn't allow for losses due to uneven

Table 2-3. Proportions by Weight to Make 1 Cubic Foot of Concrete

Maximum-Size Coarse Aggregate, in.	Air-Entrained Concrete				Concrete without Air			
	Cement, lb.	Sand, lb.	Coarse Aggregate, lb.*	Water, lb.	Cement, lb.	Sand, lb.	Coarse Aggregate, lb.*	Water, lb.
$3/8$	29†	53	46	10	29	59	46	11
$1/2$	27	46	55	10	27	53	55	11
$3/4$	25	42	65	10	25	47	65	10
1	24	39	70	9	24	45	70	10
$1\frac{1}{2}$	23	38	75	9	23	43	75	9

NOTE: Above proportions based on wet sand. See page 27 for adjustments for damp or very wet sand.
*If crushed stone is used, decrease coarse aggregate by 3 pounds and increase sand by 3 pounds.
†Metric conversion: 1 lb. = 0.454 kg; 10 lb. = 4.54 kg; 1 in. = 25 mm; 1 cu. ft. = 0.028 m³.

Table 2-4. Proportions by Volume

Maximum-Size Coarse Aggregate, in.	Air-Entrained Concrete				Concrete without Air			
	Cement	Sand	Coarse Aggregate	Water	Cement	Sand	Coarse Aggregate	Water
$3/8$	1	$2\frac{1}{4}$	$1\frac{1}{2}$	$1/2$	1	$2\frac{1}{2}$	$1\frac{1}{2}$	$1/2$
$1/2$	1	$2\frac{1}{4}$	2	$1/2$	1	$2\frac{1}{2}$	2	$1/2$
$3/4$	1	$2\frac{1}{4}$	$2\frac{1}{2}$	$1/2$	1	$2\frac{1}{2}$	$2\frac{1}{2}$	$1/2$
1	1	$2\frac{1}{4}$	$2\frac{3}{4}$	$1/2$	1	$2\frac{1}{2}$	$2\frac{3}{4}$	$1/2$
$1\frac{1}{2}$	1	$2\frac{1}{4}$	3	$1/2$	1	$2\frac{1}{2}$	3	$1/2$

NOTES:
The combined volume is approximately two-thirds of the sum of the original bulk volumes.
Above proportions based on wet sand. See page 27 for instructions on dry sand.

subgrade, spillage, and the like, so add 10 percent for such contingencies. In this case, that makes the total amount needed 22 cubic feet.

The quantities of material to buy can be calculated by multiplying the number of cubic feet of concrete (22 in this example) by the weights of materials needed for 1 cubic foot, given in table 2-3. Assuming the sidewalk will require air-entrained concrete, the quantities when using $3/4$-inch maximum-size aggregate would be:

$22 \times 25 = 550$ pounds of air-entraining cement
$22 \times 42 = 924$ pounds of sand
$22 \times 65 = 1,430$ pounds of gravel

Because a bag of cement weighs 94 pounds, you'll need to buy:

$550 \div 94 = 5.9$, or 6 bags

Aggregates are sold by the ton or by the cubic yard (27 cubic feet). Aggregates can be converted from pounds to cubic yards, or vice versa, by assuming a value of 90 pounds per cubic foot for the weight of sand and 100 pounds per cubic foot for the weight of coarse aggregate. Thus, 924 pounds of sand converts into 10.3 cubic feet (924 ÷ 90 = 10.3) or 0.4 cubic yard (10.3 ÷ 27 = 0.4); and 1,430 pounds of gravel converts into 14.3 cubic feet (1,430 ÷ 100 = 14.3) or 0.5 cubic yard (14.3 ÷ 27 = 0.5).

Measuring the Ingredients

Ingredients must be measured accurately to ensure production of uniform batches of quality concrete. Ingredients may be measured by weight or by volume; however, measurement by weight is recommended because it is more accurate and provides for greater uniformity. (It's easier to make adjustments in mix proportions when measuring by weight.) Bathroom scales are accurate enough for weighing the materials.

Each ingredient should be weighed in a separate container. Three- to five-gallon galvanized pails or plastic buckets are suitable. (Five-gallon plastic buckets are readily available from fast-food restaurants.) Remember to "zero" the scale with the empty container on it. After weighing each ingredient once, mark the level of the material inside the container. Subsequent batches may be measured by using this mark. The scale will no longer be required except to check the marks against the weight of material once or twice a day or when the moisture content of the sand has changed.

Although less accurate, measurements may be made by volume if no scale is available. For example, when using a ³/₄-inch aggregate, a 1:2¹/₄:2¹/₂:¹/₂

concrete mix from table 2-4 would be batched by measuring out 1 pail of cement, 2¹/₄ pails of sand, 2¹/₂ pails of coarse aggregate, and ¹/₂ pail of water. Take care when batching by volume not to overload the mixer; this reduces mixing efficiency.

Dry sand is rarely available for concrete work. Sand used on most jobs

Damp sand falls apart when you try to squeeze it into a ball in your hand.

Wet sand forms a ball when squeezed in your hand, but leaves no noticeable moisture on the palm.

Very wet sand, such as sand exposed to a recent rain, forms a ball if squeezed in your hand and leaves moisture on the palm.

contains some moisture, which must be accounted for as part of the mixing water. Sand with moisture is classified as damp sand, wet sand, or very wet sand. Differences between them are shown in the accompanying photos.

The proportions given in table 2-3 are based on wet sand, which is the condition of sand usually available. If you are using damp sand, decrease the quantity of sand given by 1 pound and increase the water by 1 pound. If your sand is very wet, increase the quantity of sand by 1 pound and decrease the water by 1 pound.

The proportions given in table 2-4 also are based on wet sand, but measurement by volume involves too many inaccuracies to justify making corrections for the moisture in damp or very wet sand. For example, moisture in sand causes an increase in volume known as *bulking*. The extent of bulking depends on the amount of moisture in the sand and its fineness. Dry sand can bulk to $1\frac{1}{4}$ times its volume when wetted. Accordingly, if you are measuring by volume, try to use wet sand.

Mixing the Ingredients

Proper mixing is an essential step in making good concrete. It is not sufficient to merely intermingle the ingredients. They must be throughly mixed so that cement paste coats every particle of fine and coarse aggregate in the mix. Mixing concrete may be done by machine or by hand.

The best way to mix concrete is with a concrete mixer. It ensures thorough mixing of the ingredients and is the only way to produce air-entrained concrete.

Small mixers from $\frac{1}{2}$- to 6-cubic-feet capacity can be rented or purchased. For extensive work around the home, it might pay to purchase a mixer. For the occasional small job, however,

it is preferable to rent a mixer from your local rental service yard.

Mixers are powered by gasoline or electricity. The gasoline-powered mixer is more versatile in that it can be operated anywhere. The electric-powered mixer is quieter and simpler to operate, but requires access to an electrical outlet.

Mixer sizes are designated according to the maximum concrete batch in cubic feet that can be mixed efficiently. This is usually 60 percent of the total volume of the mixer drum. The maximum batch size is usually shown on the identification plate attached to the mixer. For proper mixing, never load a mixer beyond its maximum batch capacity. The choice of mixer size will depend on the extent of your project and the amount of concrete that you want to handle in any one batch. Keep in mind that to mix a 1-cubic-foot batch of concrete you will have to handle 140 to 150 pounds of materials.

For best results, load the ingredients into the mixer in the following sequence.

1. With the mixer stopped, add all the coarse aggregate and half of the mixing water. If an air-entraining agent is used, mix it with this part of the mixing water.

2. Start the mixer; then add the sand, cement, and remaining water with the mixer running.

3. After all ingredients are in the mixer, continue mixing for at least 3 minutes, or until all materials are thoroughly mixed and the concrete has a uniform color.

Concrete should be placed in the forms as soon as possible after mixing. If the concrete is not placed within $1\frac{1}{2}$ hours and shows signs of stiffening, remixing for about 2 minutes may restore its workability. Discard the concrete if after remixing it is still too stiff to be workable. It is important that you never add water to concrete that has stiffened to the point where remixing will not restore its workability.

Mixing the Trial Batch   The proportions of sand and coarse aggregate recommended in tables 2-3 and 2-4 are based on typical gravel aggregates. If these proportions do not give a workable mix with your aggregates, an adjustment will be necessary. A trial batch gives you the information you need to make these adjustments.

First make a small batch of concrete using the proportions from either of the tables. Discharge a sample of concrete from the mixer into a wheelbarrow or onto a slab and examine it for stiffness and workability. If this sample is a smooth, plastic, workable mass that will place and finish well, the proportions used are correct and need no adjustment. The suitability of the sample can be judged by working the concrete with a shovel and smoothing it with a *float* or trowel. A good, workable mix is one in which the concrete is just wet enough to stick together

without crumbling. It should slide down, not run off, a shovel.

In a workable mix there is sufficient cement paste to bind the pieces of aggregate so that they will not separate when the concrete is transported and placed in the forms. There should be sufficient sand and cement paste to give clean, smooth surfaces free from rough spots (called honeycomb) when forms are stripped. In other words, there should be just enough cement paste to completely fill the spaces between the particles of aggregate and to ensure a plastic mix that finishes easily.

Adjusting the Trial Batch   If the trial batch is too wet, too stiff, too sandy, or too stony, adjust the proportions of aggregates used in the mix. If the mix is too wet, it contains too little aggregate for the amount of cement. Weigh out about 3 pounds of sand for each cubic foot of concrete in the batch. Add it to the trial batch in the mixer and mix for at least 1 minute. If the mix is still too wet, add some more sand until the desired workability is obtained. Record the total weight of added sand, but reduce the amount of

A workable mix contains the correct amount of cement paste, sand, and coarse aggregate. With light troweling, all spaces between coarse aggregate particles are filled with sand and cement paste.

This mix is too wet because it contains too little sand and coarse aggregate for the amount of cement paste. Such a mix will not be economical or durable and will have a strong tendency to crack.

water by 2 pounds for every 10 pounds of sand added to the trial batch.

If the mix is too stiff, it contains too much aggregate. Reduce the amounts of coarse aggregate in subsequent batches until the desired workability is obtained. Record the new weight of coarse aggregate, and correct the weight marks in the batch cans according to the adjusted weights. To save a trial batch that is much too stiff to place, cement and water may be added in the proportions of 1 pound of water to 2 pounds of cement. This will increase the amount of cement paste and make the concrete more workable. Never add water alone to a mix that is too stiff.

This mix contains too much sand and not enough coarse aggregate. It will place and finish easily, but will not be economical, because it will be very likely to crack.

coarse aggregate and add an equal weight of sand in the next batch. Record the new weights of sand and coarse aggregate and correct the weight marks in the batch cans according to the adjusted weights.

This mix is too stiff because it contains too much sand and coarse aggregate. It will be difficult to place and finish properly.

If the mix is too sandy, decrease the amount of sand by 2 pounds and add 2 pounds of coarse aggregate. If it is still too sandy, leave out some more sand and add an equal weight of coarse aggregate in the next batch. Record the new weights of sand and coarse aggregate and correct the weight marks in the batch cans according to the adjusted weights.

If the mix is too stony, decrease the amount of coarse aggregate by 2 pounds and add 2 pounds of sand. If it is still too stony, leave out some more

This mix contains too much coarse aggregate and not enough sand. It will be difficult to place and finish properly and will produce porous, honeycombed concrete.

Your adjusted trial batch proportions are your final mix proportions and need not be changed again for future batches, if your sand and coarse aggregate remain the same. If the moisture content of your sand changes, due to rain, for example, adjust the quantities of sand and water as explained earlier in "Measuring the Ingredients."

Hand Mixing For very small jobs where the volume of concrete required is less than a few cubic feet, it is sometimes more convenient, though less efficient, to mix by hand. Hand mixing is not vigorous enough to make air-entrained concrete, regardless of whether air-entraining cement or an air-entraining agent is used. Hand mixing, therefore, should not be used for concrete that will be exposed to freezing and thawing or deicers.

Hand mixing should be carried out on a clean, hard surface or in a mortar box to prevent contamination by mud and dirt. A concrete slab makes a good working surface. The measured quantity of sand is spread out evenly on the slab. Then the required amount of cement is dumped on the sand and evenly distributed. Mix the cement and sand thoroughly by turning with a short-handled, square shovel until you have a uniform color, free from streaks of brown and gray. (Streaks indicate that the sand and cement have not been thoroughly mixed.)

Next, spread this mixture out evenly over the slab and dump the required quantity of coarse aggregate in a layer on top. The materials are again turned by shovel until the coarse aggregate has been uniformly blended with the mixture of sand and cement. After at least three turnings, form a depression or hollow in the center of the pile and slowly add most of the water. Finally, turn all the materials in toward the center. Continue mixing while gradually adding the remaining water until the water, cement, sand, and coarse aggregate have all been thoroughly combined (see illustration 2-1).

Prepackaged Mixes Jobs small enough for hand mixing can usually be done more conveniently with prepackaged concrete mixes. Building materials suppliers, hardware stores, and even some supermarkets sell prepackaged concrete mixes. All the necessary ingredients—portland cement, dry sand, and coarse aggregate—are combined in the bag in the correct proportions. Packages are available in different weights, but the most common sizes are 40, 60, and 80 pounds, which make $1/3$, $1/2$, and $2/3$ cubic foot of concrete. All you do is add water and mix. Directions for mixing and the correct amount of water to add are given on the bag.

To ensure that you get good quality from prepackaged concrete mixes, the American Society for Testing and Materials has adopted "Specification for Packaged, Dry, Combined Materials for Mortar and Concrete" (ASTM C387). This specification covers the quality of the ingredients, the strength of concrete obtained with the ingredients, and the type of bag in which the ingredients are packaged. ASTM C387 requires that prepackaged concrete mixes meeting this specification be so identified on the bag. Therefore, to obtain a quality product, make sure the prepackaged mix you buy bears the statement on the bag that it meets ASTM C387.

If the concrete will be exposed to freezing and thawing or deicers, prepackaged mixes must be machine mixed and must be made with air-entraining cement, or an air-entraining agent must be added to the mixing water.

As pointed out above, prepackaged mixes are most convenient for the very small job requiring only a few cubic feet of concrete. However, for larger jobs up to 1 cubic yard (27 cubic feet), you would be wise to compare the cost of using prepackaged mixes with the cost of buying the separate ingredients. For jobs requiring more than 1

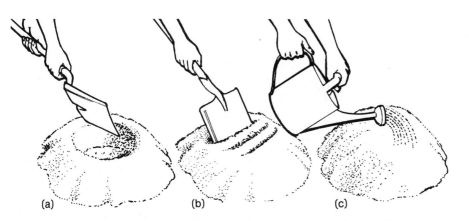

(a)   (b)   (c)

Illustration 2-1. To hand mix small batches of concrete, (a) first dry mix the cement and aggregate, form into a low mound, and make a crater in the center; (b) pour in most of the water and mix thoroughly; (c) slowly add the remaining water until you achieve the right consistency.

cubic yard, ready-mix concrete is recommended.

Cleaning the Mixer   The mixer should be thoroughly cleaned as soon as possible after you have finished using it. To clean the inside of the mixer drum, add water and a few shovels of coarse aggregate while the drum is turning. Follow this by hosing with water. The thin cement film that builds up on the exterior parts of the mixer may be removed with vinegar. Concrete that builds up inside the mixer drum requires scraping and wire brushing for removal. Heavy hammers or chisels that might tear up the drum and blades should not be used. Remove stubborn buildup with a solution of 1 part hydrochloric acid (muriatic acid) in 3 parts of water. Allow 30 minutes for penetration, then scrape with a wire brush and rinse with clear water.

CAUTION: Hydrochloric acid is hazardous and toxic and requires adequate safety precautions. Always make the solution by adding acid to water— never water to acid. Skin contact and breathing of fumes should be avoided. As a general precautionary rule, wear rubber or plastic gloves and chemical safety goggles. If the acid is used indoors, adequate ventilation should be provided. Follow the storage and handling precautions stated on the label of the acid container.

Dry the mixer drum thoroughly to prevent rusting and store the mixer with the opening of the drum pointing down. Do not apply oil to the inside of the drum unless the mixer is to be stored for an extended period of time. Thoroughly wipe off the oil before using the mixer again because it may adversely affect the quality of the concrete.

NOTE: When cleaning up mixing tools, mixers, or the work site, the federal Environmental Protection Agency and state agencies forbid discharging untreated wash water into the nation's waterways. Instead of allowing wash water to drain down a storm drain, wash out tools on a cleared area of the work site.

CHAPTER 3

Site Preparation and Formwork

The best time to plan and build a concrete drive, walk, patio, steps, or foundation is early in the construction season before the hot days of summer. As with any successful home project, a concrete project requires careful planning. A job done right will increase the value of your home and serve you well for many years. So take extra time for planning: it will help the job go more smoothly and help ensure good results.

Start by looking at a variety of different design ideas. Good sources to consult are books and magazines from your local library and brochures available at your local building supply company. Once you spot an idea you like, get a pencil and some paper and sketch out your version of it.

As you work out the details of your design, you may want to consult an architect for some expert advice. For a small fee, an architect may be able to point out options that you might never figure out on your own. Whether you consult an expert or not, once you've completed your planning, you'll need to check with your local government for approval and permits (see box).

Design and Layout: Driveways

Driveways for single-car garages or carports are usually 10 to 14 feet wide, with a 14-foot minimum width for curving drives. In any case, a driveway should be at least 36 inches wider than the widest vehicle it will serve. This extra width allows ample room on both sides of a vehicle for passengers to enter and exit comfortably. Long driveway approaches to two-car garages may be single-car width, but must be widened near the garage to provide access to both stalls. Short driveways for two-car garages should be about 16 to 24 feet wide.

The proper driveway thickness depends primarily on the weight of the vehicles that will use it. For passenger cars, 4 inches is sufficient; but, if an occasional heavy truck uses the

(a) Contact of undercarriage with driveway

(b) Contact of rear bumper with street

1¾'

12'

(c) Maximum grade should not exceed 14% (1¾" per foot).

Illustration 3-1. The driveway slope should be carefully planned to avoid damage to your vehicles and those of visitors.

Building Regulations

Before any construction begins on a driveway, sidewalk, patio, steps, foundation, or any other major home improvement, you must check with the local city or county building department. Most communities require a building permit to ensure that work is done in accordance with state and local building codes, and laws regulating building methods vary with localities. Building permits are especially important for driveways and sidewalks that cross a public way (the strip of land on either side of a street extending from property line to curb and reserved for public sidewalks, grass, and trees). Some cities may regulate the type of concrete used on various projects, and some will even set sidewalk grades and styles when a sidewalk permit is obtained.

In addition to issuing permits, local building authorities can tell you where you can get more information about the project you're undertaking. They can also advise you on which contractors in your area are reputable and which ones are not so reputable. Thus, you should view your local building authorities as valuable resources and find ways to get a good return on the cost of your permits.

driveway, a thickness of 5 or 6 inches is recommended.

If the garage is considerably above or below street level and is located near the street, the driveway grade may be critical. A grade of 14 percent (1¾ inches vertical rise for each running foot) is the maximum recommended. Also, the change in grade should be gradual to avoid scraping the car's bumper or underside. The most critical point occurs when the rear wheels are in the gutter as a vehicle approaches a driveway from the street (see illustration 3-1).

The driveway should slope slightly so that water will drain quickly after a rain or washing. A slope of ¼ inch per running foot is recommended. The

Driveways for single-car garages should be a minimum of 10-feet wide. Those for multiple-car garages should be appropriately wider.

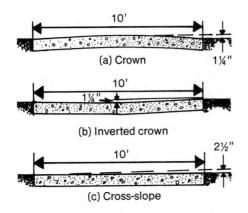

Illustration 3-2. There are three methods of obtaining side drainage for driveways with no slope.

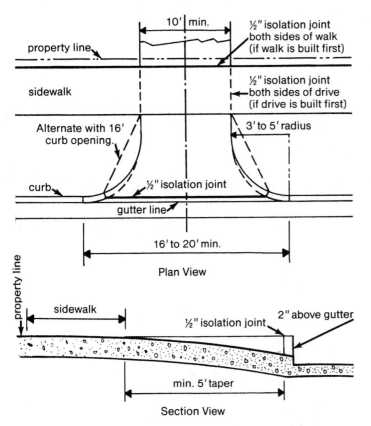

Illustration 3-3. Driveways must be carefully planned, keeping street, sidewalk, and slope in mind. Here is a typical driveway entrance.

direction of slope depends on local conditions, but usually it should be toward the street. A crown or cross-slope may be used for drainage instead (see illustration 3-2).

The part of a driveway between the street and public sidewalk is usually controlled by the local municipality; so, consult building officials when a driveway is built after the street, curbs, and public walks are in place. If curb and gutter have not been installed, it is advisable to end the driveway temporarily at the public sidewalk or property line. An entry of gravel or crushed stone can be used until curbs and gutter are built. At that time, the drive entrance can be completed to meet local requirements. If the driveway is built before the public walk, it should meet the proposed sidewalk grade and drop to meet the gutter, if no curb is planned. A typical driveway entrance is shown in illustration 3-3.

While planning the project, consider elements that can make a driveway a beautiful approach to a home rather than just a pathway to the garage. For example, you may want to include one or more paved areas for such things as offstreet parking, turning vehicles around for safe head-on street entry, and game playing.

Sidewalks

Private walks leading to the front entrance of a home should be 3 to 4 feet wide. Service walks connecting the back entrance may be 2 to 3 feet wide. The width of public sidewalks is set by local building codes. Generally, they are made wide enough to allow two people walking abreast to pass a third person without crowding. The width will vary with the amount of pedestrian traffic expected.

Walks adjacent to private properties are generally controlled by the municipality. These walks perform a valuable service by enhancing the value of the private properties. Accordingly, it has been traditional for municipalities to charge abutting properties for all or part of the cost of construction and maintenance of public walks. Check with the local government to find out who bears responsibility for these walks in your area.

In commercial or business areas walks are generally 5 or 6 inches thick. Private residential walks are not required to meet these standards, but should not be less than 4 inches thick.

It is customary to slope walks $1/4$ inch per foot of width for drainage. Where walks abut curbs or buildings, the slope should be toward the curb and away from the building over the full width of the walk. In some areas where side drainage permits, walks built with a crown or slope from center to edge are desirable. Certain conditions may require that a slope other than $1/4$ inch per foot be used—for example, where a new walk meets an existing driveway or alley.

Patios

A carefully planned patio is a valuable extension of the living and entertainment area of any home. Patio planning should account for a number of factors such as the view, climate, traffic flow to the house and kitchen, weather and insect protection, privacy, and outdoor cooking and entertaining. Location of the patio will be determined by the lot size and how the house is set on the lot. If the lot has a beautiful view of the city or surrounding countryside, the patio can be located to take advantage of the view.

Outdoor living must be oriented to the sun and general climate of the region. A south-facing patio is never deserted by the sun. It dries fast after rains and warms readily in winter months. Patios exposed to the west

More than just a beautiful approach to a home, this driveway has ample room for guest parking and turnaround area for safe head-on street entry.

Concrete sidewalks placed back from the street provide safety in your neighborhood.

are likely to be very hot in the afternoon and cool and damp in the morning. They are not as pleasant during the winter months as patios with southern exposures. In hot climates, patios facing east are better than those facing west because they cool off in the afternoon. Even better for hot climates are north-facing patios, which never receive direct sun.

The patio should be designed as part of the house. Consideration of traf-

This patio was designed level with the floor of the house and convenient to the kitchen and living room.

fic flow from the house to the patio and location of doors and windows in relation to the patio will help bring the outdoors indoors and vice versa. If the floor of the house is a few steps above ground level, you may want to build the patio slab up to the level of the house floor.

If insects are a serious problem, consider enclosing part of the patio with screening. Also, the outdoor living season can be prolonged with a roof for sun and rain protection over all or part of the patio.

The distance to the nearest neighbor usually determines the type of privacy screen needed for a patio. Options include walls, fences, or other visual barriers, such as shrubbery or growing vines. Concrete masonry screen walls are very decorative and ideally suited for this purpose.

Entertaining, cooking, and dining outdoors are among the most enjoyable uses of a patio. The patio should be spacious if considerable entertaining is anticipated. A good rule of thumb is to make the patio larger than the largest room in the house. The food-serving area should be convenient to the kitchen. Also, the location of any

future barbecue should be carefully considered.

The shape of the patio is limited only by the imagination. Square and rectangular patios are commonly used, but any shape can be built with concrete. A curved or free-form patio can be particularly attractive, especially when it complements the contour of the lawn and is accented with proper plantings. *Cast-in-place* concrete patios should be a minimum of 4 inches thick and built with a 1/4-inch-per-foot slope for drainage away from the house.

Preparing the Subgrade

The first step in building a walk, drive, or patio is preparation of the subgrade. This is a most essential step for building any concrete slab on ground. Serious cracks, slab settlement, and structural failure can very often be traced to a poor subgrade. The subgrade should be uniform, reasonably hard, free from foreign matter, and well drained.

Remove all organic matter—such as grass, sod, and roots—from the site and grade the ground. Dig out any soft or mucky spots. Fill them with soil similar to the rest of the subgrade or with granular material such as sand,

A concrete patio can be built in any shape, and free form is one of the most popular.

gravel, crushed stone, or slag. Next, thoroughly compact the subgrade. Loosen and tamp hard spots to provide the same uniform support as the rest of the subgrade. Make sure all fill materials are uniform, free of vegetable matter, large lumps or stones, and ice or frozen soil.

Use granular fills of sand, gravel, crushed stone, or slag to bring the site to uniform bearing and final grade. Compact these fills in layers not more than 4 inches thick. Extend the fill at least 1 foot beyond the slab edge to prevent undercutting during rains.

Special preparation should be given to poorly drained subgrades that are water soaked most of the time. This type of subgrade should be covered with 4 to 6 inches of granular fill. To prevent the collection of water under the slab, the bottom of this granular fill must not be lower than the adjacent finished grade. Unless fill mate-

When preparing the subgrade for a walk, drive, or patio, you need to first remove all sod, roots, and debris.

rial is well compacted, it is advisable to leave the subgrade undisturbed. Undisturbed soil supports a concrete slab better than soil that has been dug out and replaced with poorly compacted fill. Subgrade *compaction* can be done with hand tampers or a lawn roller. For a small job, hand tampers may be used, but for large-volume work such as a large driveway, sidewalk, or patio project, mechanical rollers or vibratory compactors are strongly recommended. You can hire a contractor to operate a mechanical roller. Small

Use sand or other granular fill to bring the site to uniform grade.

Hand tampers are suitable for small jobs. For best subgrade compaction, sand or gravel fill should be damp.

Vibratory compactors are practical for all but the smallest jobs.

Immediately before concreting, dampen the compacted subgrade by spraying with water.

vibratory plate compactors may be rented from an equipment rental.

The subgrade should be in a uniformly moist condition at the time concrete is placed. If necessary, dampen it by spraying with water. However, no free water should be standing on the subgrade, nor should there be any muddy or soft spots when concrete is being placed.

Formwork

Forms for drives, walks, and patios may be lumber or metal, braced by wooden or steel stakes driven into the ground. All forms should be straight, free from warping, and of sufficient strength to resist concrete pressure without bulging. For a 4-inch-thick slab, 1×4 or 2×4 lumber may be used. A 5-inch slab can be formed with 2×4s, but 2×6s are preferable. Slabs of 6-inch thickness require at least 2×6 forms. Lumber used for formwork must be smooth and of uniform thickness. Because 1×4s and 2×4s are actually only $3\frac{1}{2}$ inches in width and because slabs are typically thicker than $3\frac{1}{2}$ inches, the final grade should be slightly lower than the bottom of the form when dressed 1×4s or 2×4s are used for forming 4-inch slabs. The same applies when using 2×4s for 5-inch slabs and 2×6s for 6-inch slabs. A little *back-filling* outside the forms prevents the concrete from running under them (see illustration 3-4).

Wooden stakes are made from 1×2, 1×4, 2×2, or 2×4 lumber. They may be hand cut or purchased precut. Space stakes at 4-foot intervals for 2-inch-thick formwork. With 1-inch lumber, space the stakes more closely to prevent bulging. A maximum interval of 2 to 3 feet is suggested.

Reusable steel stakes are available in various lengths and styles for use with wooden forms. They have closely spaced nailing holes and are easier to drive and pull than wooden stakes. Because they are reusable, steel stakes can save you a lot of time and money when you do a lot of concrete work.

Setting forms to proper line and grade is normally accomplished by use of a string line as we show in a series of photos here. When making a level slab or a slab with a slope, use a straightedge and a level to assure accuracy in setting forms (see illustration 3-5). Stake and brace the forms firmly to keep them in horizontal and vertical alignment.

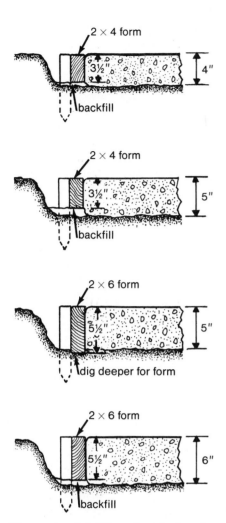

Illustration 3-4. The subgrade must be carefully fine graded to ensure proper slab thickness when using dressed lumber for formwork.

For ease in placing and *finishing* concrete, drive all stakes slightly below the top of the forms. Wooden stakes can be sawed off flush first. All stakes must be driven straight and true if forms are to be plumb. For easy stripping, use *double-headed nails* driven through the stake into the form and not vice versa.

Forming Curves   Horizontal curves may be formed with 1-inch lumber, 1/4- to 1/2-inch-thick plywood, hardboard, or sheet metal. Short-radius curves are easily formed by bending plywood with the grain vertical. Bending 2-inch-thick wooden forms is accomplished by using stakes for gentle curves and saw kerfs for shorter-radius curves (see illustration 3-6). Wet lumber is easier to bend than dry lumber.

Gentle vertical curves can sometimes be formed by bending a 2 × 4 during staking. When the change in slope is sharper, a series of short lengths of the form material may be used. The curve is laid out with a string line tied to temporary stakes. The line is adjusted up or down on the stakes to give a smooth curve, then short lengths of forming are set to the string line and securely staked. You can also cut saw kerfs into the form material to make it bend more easily into vertical curves. To hold forms properly in vertical or horizontal curves, set stakes closer together than you would on straight runs.

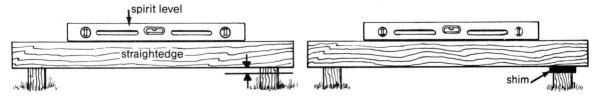

Illustration 3-5. To set the slope of a slab, set stakes level on either side then pound the stake in on one side the appropriate amount (for example, 1/2 inch). Level the straightedge again with a wood shim, then repeat the process for the rest of the slab.

Using a level, mark stakes with proper grade and string a line tightly from stake to stake. Drive intermediate stakes in line with the string. Check the distance from string to subgrade to make sure there is enough depth to place the forms.

Attach the forms to stakes with nails, which should be driven through the stake and into the form as shown here. Hold the form tightly against the stake and level with the string by foot pressure. Double-headed nails are recommended for easy form stripping.

Set the forms so that their tops are level with the string line. If there is insufficient room under the string for the form, dig out the subgrade.

Permanent Forms Wooden side forms and divider strips may be left in place permanently for decorative purposes and to serve as *control joints*. Such forms are usually made of 1×4 or 2×4 redwood, cypress, cedar, or treated wood. Treated wood requires no special preparation; but, redwood, cypress, and cedar should be primed with a clear wood sealer before use. It's also a good practice to mask the top surfaces with tape to protect them from abrasion and staining during concrete placing and finishing.

Make neat butt or miter joints at the corners and join intersecting strips with neat end-to-end butt joints reinforced with stakes placed on the outside of the forms. Drive 16d galvanized nails at 16-inch intervals horizontally at midheight through stay-in-place forms to help anchor them to the concrete after it is placed. Interior *divider strips* should have nail anchors similarly spaced but driven from alternate sides of the board (as shown in illustration 3-7). Drive all nail heads flush with the forms, but never drive nails through the top of permanent forms. All stakes that are to remain in place permanently should be driven down or cut off 2 inches below the surface of the concrete, so they will not become a safety hazard.

Forming curves is easily done with 1 × 4 lumber. Inside stakes are pulled out after outside stakes are nailed firmly to the forms.

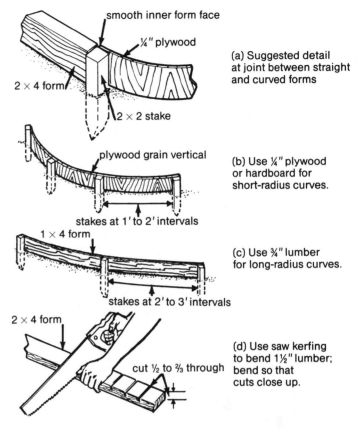

(a) Suggested detail at joint between straight and curved forms

(b) Use ¼" plywood or hardboard for short-radius curves.

(c) Use ¾" lumber for long-radius curves.

(d) Use saw kerfing to bend 1½" lumber; bend so that cuts close up.

Illustration 3-6. Horizontal curves are easy to form using the methods shown here.

To preserve their color and to protect the forms from abrasion and staining by concrete, use masking tape to cover the top surfaces of divider strips and outside forms that are to remain in place permanently.

Final Check Before concreting, give all forms a final check for trueness to grade and proper slope for drainage. Check the subgrade with a wooden template or a string line to ensure correct slab thickness and a smooth sub-

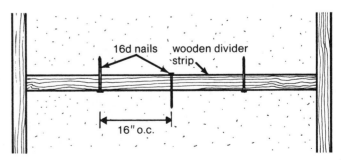

Illustration 3-7. When using permanent forms for a concrete slab, drive 16d galvanized nails through the form 16 inches as on center. On interior forms, alternate the direction of the nails.

grade. Make sure joints have been planned and their location marked on the forms. (See the detailed discussion of jointing concrete at the end of this chapter.) Finally, dampen forms with water or oil them for easier removal. Motor oil may be used for small jobs, but a regular form release agent applied with hand-spray equipment is preferred for volume work.

Table 3-1.	**Recommended Values for Stepped Ramps**			
	Paired Risers		**Single Risers**	
	Min.	Max.	Min.	Max.
Riser height	4″	6″	4	6″
Tread length	3′ 0″	8′ 0″	5′ 6″	5′ 6″
Tread slope	1/8″/ft.	1/4″/ft.	1/8″/ft.	1/4″/ft.
Overall ramp slope	2 1/8″/ft.	3 1/4″/ft.	15/16″/ft.	1 7/16″/ft.

NOTE: Recommended values given provide one or three easy paces between paired risers and two easy paces between single risers.

A wooden template is used to check the form for proper slab thickness and smooth subgrade.

Steps

Steps at entranceways must conform to the provisions of local building codes. These codes establish critical dimensions such as width, height of flights without landings, size of landings, size of risers and treads, and the relationship between riser and tread size.

Steps for private homes are usually 48 inches wide. Some codes allow 30- and 36-inch widths; however, steps should be at least as wide as the door and walk they serve. A landing is desirable to divide flights of more than 5 feet and it should be no shorter in direction of travel than 3 feet. The top landing should be no more than 7 1/2 inches below the door threshold.

For flights less than 30 inches high, maximum step rise is usually 7 1/2 inches and minimum tread width is 11 inches. For higher flights, step rise may be limited to 6 inches with a minimum tread width of 12 inches. Choice of riser and tread size should depend on how the steps are to be used. For aesthetic reasons, steps with risers as low as 4 inches and treads as wide as 19 inches are often built. On long, sloping approaches, a stepped ramp can be used. Two versions are shown in illustration 3-8, and table 3-1 provides a range of values to guide the construction of each type.

Many studies have been made to find the best combination of riser and tread for optimum comfort and safety. One study concludes that the sum of riser and tread should equal 17 1/2 inches. This is a good combination for most steps; however, more generous steps may be desirable in leisure areas such as patios, gardens, and terraces, or for aesthetic purposes. As a general rule, the closer the climbing of steps comes to the normal walking stride, the safer and easier it is for all ages.

For safety there should be no variation in the height of risers and the width of treads in any one flight.

Footings

Footings for steps should be placed at least 2 feet deep in firm, undisturbed soil and, in areas where freezing occurs, 6 inches below the prevailing frost line. Steps with more than two risers and two treads should be supported on concrete blocks, concrete walls, or piers at least 6 inches thick, or should be cantilevered from the main foundation walls. Steps should be securely tied to foundation walls with *anchor bolts* or tie-rods. On new construction, step footings may be cast integrally with foundation walls.

To prevent new steps from sinking when they are added to an existing building, dig two or more 6- to 8-inch-diameter postholes beneath the bottom tread and fill them with concrete. The holes should extend to the depth indicated above for footings. The top step or landing should be tied into the foundation wall with two or more metal anchors.

Formwork

Forms for steps must be rigidly braced to prevent bulging and tight to prevent leaking. They must be built for easy stripping without damage to the concrete. Forming materials should be straight, true, and free from imperfections that would be visible in the hardened concrete. Wood containing protruding or missing knots, bent nails, or other blemishes makes finishing more difficult and the final appearance less attractive.

Before proceeding further, it is important to outline the two methods of finishing steps, because the method used affects how the forms are built.

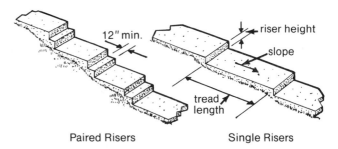

Paired Risers   Single Risers

Illustration 3-8. When building stepped ramps for sloping yards, use the guidelines given in table 3-1 to get the right dimensions for treads and risers.

In these concrete and stone steps, low-profile risers, wide treads, and intermediate landings are combined to complement the sloping landscape.

Steps to this building will be supported on 6-inch-thick concrete walls. Broken concrete used as fill will be covered with well-compacted sand before the steps are cast.

One way to provide a solid footing for steps is to integrally cast them with the basement walls.

In one method, stripping of riser forms begins 30 minutes to several hours after placing concrete, depending on weather conditions and how fast the concrete sets. At this point, the steps should be strong enough to support their own weight. Rough edges and formed surfaces are rubbed with a float to bring mortar to the surface, and additional mortar can be plastered and troweled onto the surface in a thin layer. After some stiffening, the mortar is troweled and brushed.

In the other method, treads and landings are finished soon after placing, and all forms are left in place for several days until curing is completed. After stripping, any necessary chipping, hand stoning, or patching is done, and riser and sidewalls are given a *grout* cleandown to fill all voids and remove any mortar left on the surface.

Step sidewalls can be formed with plywood or hardboard panels, or with board materials, as shown in illustration 3-9. The forms may need to be reinforced with 2 × 4 bracing to resist sideways thrust when the concrete is placed. On wide steps, riser supports and braces should be used roughly every 4 feet on center, as shown in illustration 3-10. When using panels, riser forms must be cut to fit inside dimensions.

Installation of riser forms should start at the top to eliminate unnecessary traffic on previously placed risers. Risers should be checked with a hand level and positioned to allow a ¼-inch slope on each tread for drainage. A bevel on the lower edge of the riser form permits finishing the full width of the tread. Riser forms are sometimes tilted or battered about 1 inch to increase the width of the treads. (The bevel and tilt on riser forms is shown in illustration 3-9.) Riser forms must be securely fastened to the sidewall forms and well braced to prevent bending under pressure of the concrete. Wooden cleats are the best way to attach riser forms to plywood sidewall forms. When building steps between masonry or concrete walls, use wooden wedges to help hold the risers in place.

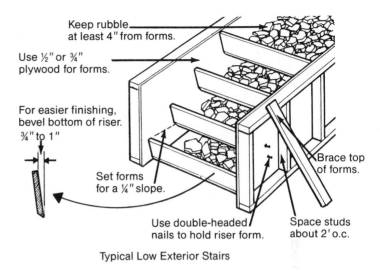

Keep rubble at least 4" from forms.

Use ½" or ¾" plywood for forms.

For easier finishing, bevel bottom of riser. ¾" to 1"

Set forms for a ¼" slope.

Use double-headed nails to hold riser form.

Brace top of forms.

Space studs about 2' o.c.

Typical Low Exterior Stairs

Illustration 3-9. When building step forms, keep in mind that treads need to slope slightly forward for drainage. It's also a good idea to angle the risers in and bevel the bottom of the riser forms for ease of finishing.

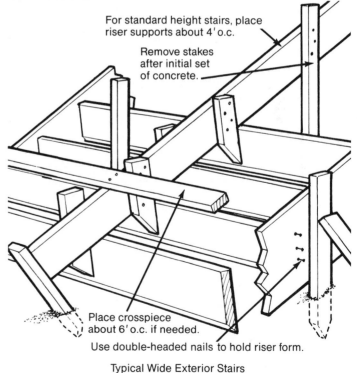

For standard height stairs, place riser supports about 4' o.c.

Remove stakes after initial set of concrete.

Place crosspiece about 6' o.c. if needed.

Use double-headed nails to hold riser form.

Typical Wide Exterior Stairs

Illustration 3-10. Riser forms on wide exterior steps need supports placed every 4 feet on center.

Risers should be level from left to right and positioned so that the back of each tread will be ¼ inch higher than the front.

Wooden wedges are used to hold riser forms between concrete or masonry walls.

During forming, provision should be made for any recesses needed for attaching ironwork or railings. Also, an isolation joint is required where the top tread or landing meets the building (see "Jointing," at the end of this chapter). If this is not done, the concrete may *bond* to the wall and someday cause a crack.

Fill Materials Brick, stone, or broken concrete can be used inside the formwork as fill to reduce the amount of concrete needed. Soil or granular fill placed in well-compacted layers may also be used. There should be no fill material any closer than 4 inches from the face of any form.

Falsework (inside forms) may be used to reduce the concrete required in large steps. When falsework of any absorbent material such as wood is to remain in place permanently, it should be no closer than 6 inches from the face of any outside form. If falsework is protected with a moisture-proof membrane such as plastic sheeting, the con-

crete thickness may be reduced to 4 inches. Improperly protected wooden falsework absorbs water from the concrete and swells. This swelling, coupled with normal drying shrinkage in concrete, can cause severe cracking.

Basements, Slab Floors, and Foundations

Concrete is the material most widely used for house foundations and for residential basement floors and walls. Concrete is commonly found in basements because a properly constructed concrete basement is relatively low in cost, strong, durable, dry, and fire safe. However, methods that are effective in sealing out both water and water vapor must be undertaken early in construction, before the basement floor is placed or backfill is pushed against the foundation wall. Warm in winter, cool in summer, concrete basements can be comfortable living spaces for games, hobbies, laundry, and storage.

Site Considerations

The highest area on a given plot of land should be selected for locating a house, particularly when the land is near a body of water or a swamp. The house should be so placed on a hillside site that runoff water will not dam up against the uphill face of the basement wall. Diagonal placement of a house on such a site can often prevent this from happening.

If experience in the area indicates that a high groundwater table or springs may be present, a test boring should be made. The boring can be made with simple auger tools to a depth of approximately 15 feet, the depth of boring usually necessary to locate groundwater that can affect the performance of concrete basements and slabs-on-grade for residential construction. The height of any standing water in the hole will indicate the elevation of the groundwater. Also, the boring will show the type of soil present such as gravel, sand, silt, or clay.

Grading and Surface Drainage

Basement excavation and subsequent backfilling normally results in some settlement of backfill adjacent to the footing and foundation walls. Backfill should be brought to approximately the same density and moisture content as that of adjacent soils. This can be done by returning it to the trench in shallow layers and properly compacting each layer.

Grading of the surface should result in a fall of 1/2 to 1 inch per foot for at least 8 to 10 feet away from the foundation walls after the backfill has settled. On hillside sites, a cutoff drain may be necessary to divert water away from the foundation on the high side of the house. On a low site, the house should be constructed on high foundations and fill brought up around the walls to allow water to flow away on all sides.

Subgrade Conditions

The strongest soils are sands and gravels, which provide excellent bearing support for all ordinary footings and foundations. In contrast, some silts and clays are not strong and, when wet, will compress under relatively small loads. Gravel and sand are coarse-grained soils with fair to excellent value as foundation support. Silt and clay are fine-grained soils with poor to fair value as foundation support. The latter can be expansive and cause future problems related to building movement such as heaving of floors and cracking of walls. Even in very dense clays, the groundwater can rise 11 to 12 feet above the level of the water table and result in serious moisture problems in basements not adequately drained and waterproofed.

Clay soils classed as *expansive* contract when they dry and swell when wet. These soils may swell as much as five percent during a wet season and produce uplift forces sufficient to crack footings, walls, and floors. Where expansive soils are encountered, it's best to use foundation designs that concentrate the weight of the structure on a smaller area such as pad footings or drilled piers (see illustration 3-11). In all cases, a floating plain concrete floor slab is used.

Basement Drains

In addition to the site preparations discussed earlier for basements and foundations, floor drains must be located and placed. Floor drains in the basement tie into the same sewer line as the main and secondary stack. These lines should all be installed in trenches dug to a depth sufficient enough to provide adequate drainage to the sewer line. The lines should rest on undisturbed earth in order to provide the best support. If fill is required under the pipe, use sand, not loose dirt.

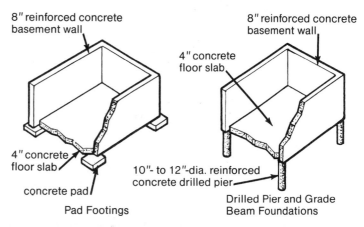

8" reinforced concrete basement wall

8" reinforced concrete basement wall

4" concrete floor slab

4" concrete floor slab

concrete pad

10"- to 12"-dia. reinforced concrete drilled pier

Pad Footings

Drilled Pier and Grade Beam Foundations

Illustration 3-11. Footing and foundation designs for expansive clay soil conditions include the use of pad footings or drilled piers with grade beam foundation walls.

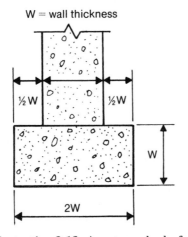

Illustration 3-12. As a general rule, footings should be the same thickness as and twice the width of the foundation walls that will sit on them.

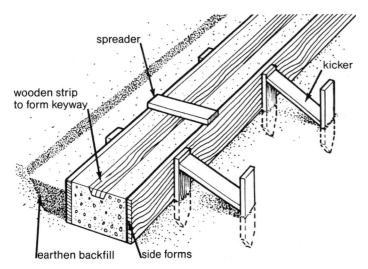

Illustration 3-13. For noncompact soils, use the above forming technique for footings. If the soil is compact, the earth will serve as a form if the footing is carefully dug.

If the sewer line is above the floor of the basement, you'll need to install a sump pump. A sump pump fits inside a 24-inch-diameter collection pit that's about 2 to 3 feet deep. A clay sewer tile works well for this. Invert the tile with the flange end down and line the inside with a 2- to 3-inch bed of gravel. Lay a piece of concrete, such as a patio block, on the gravel, then set the sump pump on the concrete pad.

Slab Foundations

Foundations for small structures may be made using a simple slab. A 4-inch slab can support a one story structure. If the subgrade is hard, dry, and frost free, you can cast the slab directly on the ground. If the soil is dark and contains humus, remove about 4 inches of soil and add 4 inches of crushed stone or gravel, and extend the gravel 12 inches beyond the planned slab.

Setting the forms is done using the same methods as described above for drives and walks. Drainage for foundation slabs should be $1/8$ inch per foot and slabs should slope toward the edges or to openings such as a garage door.

Footings

Foundations not resting on pads or piers are constructed on footings. Footings are constructed directly on a prepared subgrade with the base of the footing below the frost line or at a minimum depth of 18 inches. If there is bedrock at any level, it can be scrubbed clean and the footing cast directly on it.

To assure stability, footings should be 8 to 12 inches thick and at least twice the width of the wall constructed on them (see illustration 3-12). Forms for the sides of footings are made and set in the same manner as for slabs described above except that $3/4$-inch plywood may be required as a form material to compensate for the additional height.

In good cohesive soils that stand up well, excavation down to the top of

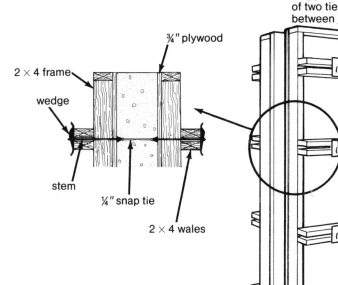

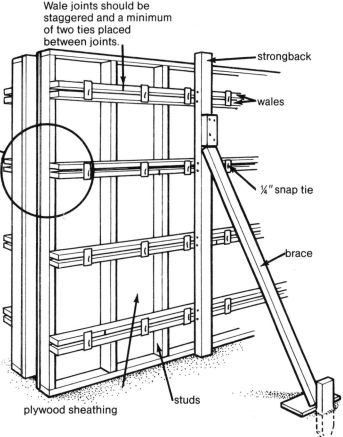

Wale joints should be
staggered and a minimum
of two ties placed
between joints.

strongback

wales

¼" snap tie

brace

studs

plywood sheathing

¾" plywood

2 × 4 frame

wedge

stem

¼" snap tie

2 × 4 wales

Illustration 3-14. Forms for casting basement walls can be made of plywood and 2 × 4s and are braced securely. Braced strongbacks should be used every 16 feet on each side of the wall.

the footing is done with a trencher or backhoe, and the footing profile is cut by hand or by specialized mechanical equipment to the exact size. If the soil is porous or noncohesive, the rough excavation proceeds to the bottom of the footing, and wood is used to form the sides of the footing. Because footings are generally shallow, lateral pressure from the fresh concrete is relatively small, and the required bracing is simple. Forms for one side of the footing are set to line and grade, and staked into position. After one side is set, the other side is aligned by spreaders and a spirit level held in position by stakes placed about every 6 feet. If the holding power of the stakes is poor because of ground conditions, the forms can be braced with a kicker to a second stake or with earthen backfill against the forms, as shown in illustration 3-13. If the earth is firm, concrete may be poured directly into the trench but care must be taken to make sure the footing is level.

Basement Walls

Forms for basement walls can be constructed either from prefabricated panels or from conventional plywood sheathing, studs, *wales,* and bracing (see illustration 3-14). As indicated by the illustration, formwork for basement walls is a complex operation so you may choose to have this operation done by a reputable concrete contractor.

It is convenient to lay prefabricated form panels close to their final position around the perimeter of the excavation and spray them with form release agent. Before setting the foundation forms, the footing should be swept clean of dirt and debris.

Concrete Reinforcement

Early builders used concrete to carry compressive loads. The arch was a common structural shape because arches can be built to include only compressive stresses, and arched doorways and window openings can still be seen on many structures, old and new.

This early use of concrete was wise but limited in its application. Concrete is extremely strong in compression. A load of several thousand pounds is required to crush a 1-square-inch section, as noted earlier. On the other hand, concrete is weak in tension. A square inch pulls apart under a load of only a few hundred pounds. In short, the compressive strength of concrete is about 10 times its tensile strength (see illustration 3-15).

When reinforcement is used in concrete, the reinforcement withstands tensile stresses. Depending on how much reinforcement is used, the tensile load can be made to equal the compressive load (see illustration 3-16).

Types of Reinforcement

Through the years many materials have been tried as reinforcement in concrete, but today steel is universally accepted and used. One important characteristic of steel is that it has nearly the same temperature expansion characteristics as concrete.

Reinforcing steel can be purchased in the form of reinforcing bars (called *rebar* for short) or as welded wire fabric (called mesh). Rebar may be either smooth or deformed. Smooth bars are usually of small diameter. *Deformed*

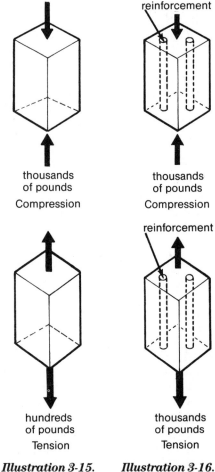

Illustration 3-15. Compression and tension are two different properties of concrete. It takes a great deal of pressure to crush concrete, but comparatively little to pull it apart.

Illustration 3-16. When reinforcing material is used in concrete, compressive and tensile strengths become the same.

bars have luglike ridges that increase the bond of the concrete to the steel. Bars come in standard sizes and are properly designated by number. These numbers describe the diameter of bar in terms of eights of an inch. A #4 bar, for example, is ⁴/₈ inch in diameter, a #5 bar is ⁵/₈ inch, and so on.

There are 11 sizes of deformed bars: #3 to #11 inclusive, plus #14 and #18. The size of bar to use depends on the amount of tensile force the concrete will need to carry. Bars are usually supplied by the mill in 20-, 40-, or 60-foot lengths and may be purchased through a building materials supplier.

Wire mesh is made in many styles and sizes, but the most common style has wires spaced at 6-inch centers both ways. Six-gauge, 8-gauge, and 10-gauge fabric is commonly available. Welded wire fabric is used for jobs requiring only relatively light reinforcement.

Using Reinforcement

Concrete expands and contracts with changes in temperature and moisture. Reinforcement is sometimes used to control cracking due to these changes. Although steel will not prevent formation of cracks, it will distribute them and make them smaller. It holds cracks tightly closed and prevents the cracked sections from moving apart. With firm, uniform subgrade support and the maximum joint spacing recommended in table 3-2, steel reinforcement is usually not necessary for slabs-on-grade.

Another application of reinforcement is when concrete is used to span any distance where there is little or no support. Examples of this are lintels for doors and windows, footings that span over pipe trenches, and anchoring steps to foundations. Reinforcement may also be used to secure piers and foundations to footings.

If the reinforcement you are using is not continuous for the length of the construction, overlap rebar at least 12 inches and mesh at least one square (see illustration 3-17). In laying out

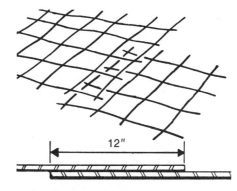

Illustration 3-17. Overlap concrete reinforcement at least 12 inches for deformed bars (more for smooth bars) and one square for mesh.

Table 3-2.	**Maximum Spacing of Control Joints in Feet**		
	Slump 4 in. to 5 in.		
Slab Thickness, in.	**Maximum-Size Aggregate Less Than ¾ in.**	**Maximum-Size Aggregate ¾ in. and Larger**	**Slump Less Than 4 in.**
4	8	10	12
5	10	13	15
6	12	15	18

the reinforcement, support the rebar on wire stilts (bar supports) made specifically for the purpose, or on pieces of broken concrete or rocks. Never use clay brick for this purpose. When in place, the reinforcement should be in the center of the form so that it will be covered with at least 1½ inches of concrete. Remove any heavy rust or dirt from the reinforcement prior to placing concrete.

Jointing

As noted earlier, concrete contracts and expands slightly under varying conditions of moisture and temperature. When newly placed, concrete occupies its largest volume. When dry and cold, it contracts to its smallest volume. These changes in volume are normal and produce stress in concrete that is restrained.

Cracks will appear where the stresses to pull concrete apart exceed its strength to hold itself together. Volume changes are an unavoidable, inherent property of concrete, so when you work with concrete, you are always confronted with the need to control stress buildup and eliminate unsightly, random cracks. Volume changes are influenced by the total water content in the mix, weather conditions, and other factors. More specifically, excess water used in the mix increases the likelihood of shrinkage cracking. Joints are used to solve the problem of random cracking of concrete by inducing the concrete to crack in straight lines at predictable locations.

Types of Joints

There are three kinds of joints used in concrete work:

• *Isolation joints* (also called *expansion joints*), which allow movement where newly placed concrete abuts existing parts of a building

• *Control joints* (also called *contraction joints*), which induce cracking at preselected locations

• *Construction joints*, which provide stopping places during construction

Isolation Joints Isolation joints usually consist of premolded strips of fiber material. For slabs-on-grade they should extend the full depth of the slab or slightly below it. These strips need not exceed $1/2$ inch in thickness (see illustration 3-18). Isolation joints are required at points of potential restraint to movement—for example, against fixed objects such as walls, columns, footings, intersections of other concrete elements such as walks and drives, and at points where slabs abut existing buildings or curbs (see illustration 3-19). Isolation joints *are not* required at regular intervals in sidewalks and driveways.

Isolation joints should be full flush with the finished surface or, better still,

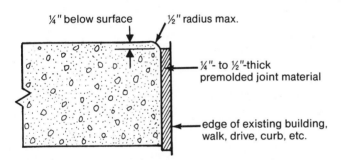

Illustration 3-18. An isolation joint is used where new concrete abuts existing concrete or a building. Use a premolded joint material made especially for this purpose.

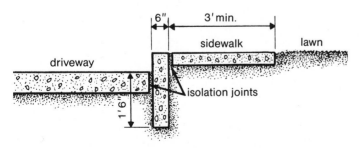

Illustration 3-19. This is an example of a typical application of isolation joints used in driveway-curb-sidewalk construction.

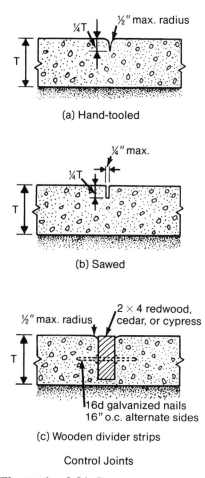

T = thickness

(a) Hand-tooled

(b) Sawed

(c) Wooden divider strips

Control Joints

Illustration 3-20. Because concrete by nature cracks, control joints determine where it will crack. Here are three different types of control joints commonly used in concrete construction.

about ¼ inch below. Joint fillers that protrude above the surface are a safety hazard. For this reason, joint materials that extrude when compressed should not be used.

Control Joints Control joints are either hand tooled into fresh concrete, sawed into hardened concrete, or formed with wooden or premolded divider strips (see illustration 3-20). The tooled, sawed, or premolded joint should extend into the slab one-fourth of the slab thickness. A joint of this depth provides a weakened section that induces a crack to occur beneath the joint where it will be inconspicuous.

In driveways, sidewalks, and patios, control joints should be spaced at intervals equal to not more than 30 times the slab thickness unless low-slump concrete of less than 4 inches is used (see table 3-2). Drives and walks wider than about 10 to 12 feet should have a longitudinal control joint down the center. If possible, the panels formed by control joints in walks, drives, and patios should be approximately square. Panels with an excessive length-to-width ratio (more than 1½:1) are likely to crack near midlength. As a general rule, the smaller the panel, the less likely random cracking will occur. All control joints should be placed as continuous, not staggered or offset, lines.

Construction Joints Construction joints are inserted where concrete placement is suspended for 30 minutes or more or along the perimeter of the day's work. Construction joints are normally located and constructed so as to act as control joints. A butt joint is satisfactory for 4-inch-thick slabs; but for thicker slabs, a tongue-and-groove joint can be used to provide load transfer across the joint and to ensure that adjoining slabs will remain level. The tongue and groove may be formed by fastening metal, wooden, or premolded key material to a wooden bulkhead. Concrete at the joint should be edged with a hand tool or sawed to match the control joints in appearance (see illustration 3-21).

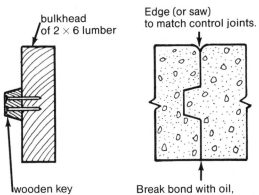

Construction Joint

Illustration 3-21. For slabs thicker than 4 inches, keyed tongue-and-groove joints are used for construction joints.

Illustrations 3-22 and 3-23 show the type and location of joints for typical residential concrete work. When planning your concrete project, plan your joint locations: they're not only critical for good concrete work, they also add an aesthetic design element that must be taken into consideration. Before you begin placing concrete, it's a good idea to measure and mark joint locations on the forms.

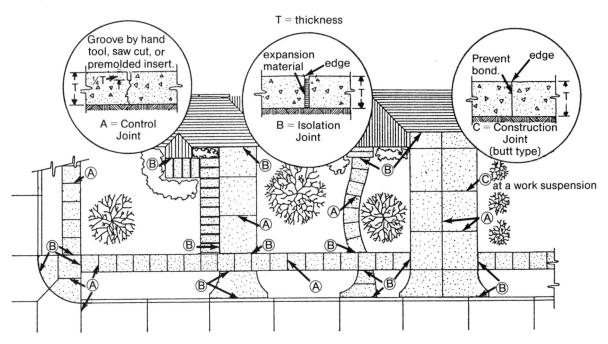

Illustration 3-22. Concrete walks and driveways should be carefully provided with joints.

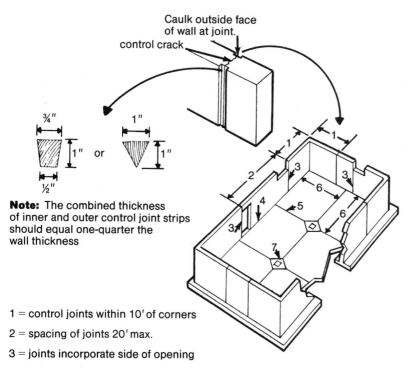

Caulk outside face of wall at joint.

control crack

¾"

1" or 1"

½"

Note: The combined thickness of inner and outer control joint strips should equal one-quarter the wall thickness

1 = control joints within 10' of corners

2 = spacing of joints 20' max.

3 = joints incorporate side of opening

4 = isolation joint between floor and wall

5 = control joint in floor slab

6 = floor slab joint spacing (see table 3-2)

7 = isolation joint around column footing*

*May be omitted if column footings are below floor level and the column is wrapped with two layers of building paper or isolation joint material to break the bond.

Illustration 3-23. Make sure you consider the design of control and isolation joints when building basement walls and foundations.

CHAPTER 4

Working with Concrete

Advance planning and preparation for delivery of concrete saves time and prevents confusion on the job. Forms and subgrade must be ready, and the manpower, tools, and materials needed for placing, finishing, and curing the concrete must be on hand. Here are some of the basic items needed for a typical job:

Wheelbarrow	Groover
Shovel	Float
Straightedge	Trowel
Bull float or	Broom
darby	Water hose
Edger	Curing materials

If concrete can be discharged directly onto the subgrade, two or three people will generally be required. One or two additional people will be needed if concrete must be wheelbarrowed from truck to subgrade. Plan to get the heavy truck mixer as near as possible to the point of placement without driving it over existing sidewalks or driveways. If the job site is especially difficult to reach, ask the ready-mix producer for suggestions before delivery of the concrete.

Placing Concrete

Preparation prior to placing concrete includes trimming, compacting, and moistening the subgrade; erecting forms; and setting the reinforcing steel and other embedded items securely in place (as discussed in the previous chapter). Moistening the subgrade is especially important in hot weather to keep the dry subgrade from drawing water out of the concrete. In cold weather the subgrade must not be frozen, and snow and ice must be removed from within the forms before concrete is placed. Where concrete is to be placed on rock, all loose material must be removed, and cut faces should be nearly vertical or horizontal rather than sloping.

Forms should be accurately set, clean, tight, adequately braced, and constructed of, or lined with, materials that will impart the desired formed-surface finish to the hardened concrete. Wooden forms should be moistened before placing concrete, otherwise they will absorb water from the concrete and swell. Forms should be made for removal with minimum damage to the concrete. With respect to wooden forms, this means avoiding too large or too many nails and using double-headed nails because they are easy to remove. Forms should also be treated with a release agent such as oil or lacquer to make removal easier.

Reinforcing steel should be clean and free of loose rust or mill scale. If you place concrete in stages, mortar stains more than a few hours old should be removed from the steel before each pour. In general, make sure all equipment that will be used to place the concrete is clean and in good working condition.

Depositing the Concrete

Concrete should be deposited as closely as possible to its final position. In slab construction, placing should start in a corner at one end of the work with each batch discharged against previously placed concrete. The concrete should not be dumped in separate piles and then leveled and worked together, nor should the concrete be deposited in big piles and then moved horizontally into final position. These practices result in segregation because mortar tends to flow ahead of coarse material.

In walls, place concrete at the ends first and work toward the center. Be careful to keep water from collecting at the ends, in corners, and along faces of forms. Concrete should be placed in horizontal layers of uniform thickness, each layer being thoroughly consolidated before the next is placed. The rate of placement should be rapid enough so that the previous layer of

When ordering ready-mix concrete, make arrangements to have the truck get as close as possible to your site. Ready-mix producers can provide extra chutes if necessary, but they may need to be ordered with the concrete.

concrete is still plastic when the next layer is placed on it. This will avoid flow lines, seams, and planes of weakness called *cold joints,* which result when fresh concrete is placed on hardened concrete. Layers should be 6- to 20-inches deep for work such as basement walls.

When casting slabs, place concrete uniformly to the full depth of the forms and as closely as possible to final position. Do not drag or flow the concrete excessively. *Overworking the mix in this manner causes an excess of water and fine material to be brought to the surface. This may cause scaling and dusting later on.* Using a square shovel, spade concrete along the forms to compact it firmly and eliminate voids and honeycomb.

Spreading and Compacting the Concrete

Spreading and spading are best done with a short-handled, square shovel. Spreading can also be done with special concrete rakes or hoe-like tools known as *come-alongs.* Ordinary rakes and hoes should not be used because they separate large pieces of gravel or stone from the mortar in the mix.

Compacting fresh concrete is a process also referred to as consolidation. Its purpose is to mold the concrete within the forms and around embedded parts and reinforcement and to eliminate voids. Compacting concrete in walls or deep forms is typically done by hand using a process known as *hand rodding,* which is done by thrusting a tamping rod or other suitable tool into the concrete. The rod should be long enough to reach the bottom of the form, and thin enough to pass between any reinforcing steel and the forms.

Spading can be used to improve the appearance of formed surfaces. A flat, spadelike tool should be repeatedly inserted and withdrawn adjacent to the form. This forces the larger coarse aggregates away from the forms and assists entrapped air bubbles in their movement toward the top surface.

The best way to consolidate concrete in walls is with an internal *vibrator,* a tool that can be rented. The vibrator is inserted the full depth of

After the concrete is deposited on the subgrade, it is spread with a short-handled, square shovel.

each lift and about 6 inches into the previous lift for about 10 seconds every 6 to 12 inches along the wall.

Placing New Concrete on Hardened Concrete

Even though best results are achieved when all concrete is placed at the same time, there are times when this is not possible. If you wish to place fresh concrete on hardened concrete, you need to take certain precautions in order to secure a well-bonded, watertight joint. The hardened concrete must be clean, moist, fairly level, and reasonably rough with some coarse aggregate particles exposed. Any laitance, soft mortar, dirt, wood chips, oil, or other foreign materials should be removed from the top surface because they will interfere with proper bond.

The surface of the hardened concrete may be prepared either before or after the concrete sets. For concrete less than 8 hours old, remove any laitance, loose particles, and dirt. For older concrete, thoroughly clean the surface of films and deposits: this may require roughening with a chipping hammer, water jet, or sandblasting.

Hardened concrete should be moistened thoroughly before new concrete is placed on it; however, its surface should be completely free of puddles or shiny spots that indicate free moisture.

Leveling and Smoothing Concrete

After concrete has been spread and compacted to fill the forms, strikeoff and bull-floating (or darbying) follow immediately. It is of utmost importance that these operations be performed before bleed water has an opportunity to collect on the surface. Concrete should not be spread over too large an area before strikeoff, nor should a large area be struck off and allowed to remain before bull-floating or darbying. *Any operation performed on the surface of a concrete slab while bleed water is present will cause serious dusting or scaling. This point cannot be overemphasized. It is the basic rule for successful finishing of concrete flatwork.*

Strikeoff

Strikeoff is the operation of removing concrete in excess of the amount required to fill the forms and bring the surface to grade. It also smooths and shapes the surface and eliminates any high or low spots. The tool used is known as a straightedge. Straightedges are made of metal or wood and may be equipped with a vibrator that compacts the concrete during strikeoff. Straight 2×4 lumber may be used; or wooden straightedges can be made especially for the purpose. The straightedge should be 12 to 18 inches longer than the width of the slab.

A vibrating straightedge gives positive control of the strikeoff operation and saves a great deal of time. The device shown is engine driven and can be found at some tool rental dealers.

Striking off may be done with a straight piece of 2 × 4 lumber.

You can make a straightedge especially for striking off concrete. This straightedge has hand grips on it for better control while screeding. Place and strike off concrete in small areas at a time.

Concrete is struck off by sliding the straightedge back and forth along the tops of the forms with a sawlike motion. Keep a small amount of concrete ahead of the straightedge to fill in low spots. As the straightedge advances forward, it should be tilted slightly in the direction of travel to obtain a cutting edge. Work about a 30-inch-long section of the slab in one pass, then go back over it in a second pass to remove any remaining bumps or low spots. During the second pass the straightedge should be tilted slightly in the opposite direction.

Bull-Floating and Darbying

Bull-floating or darbying immediately follows strikeoff. The purpose of either of these operations is the same: to level ridges and fill voids left by the straightedge and to embed all particles of coarse aggregate slightly below the surface. A bull float is used for areas too large to reach with a darby.

Bull floats are large, long-handled floats made of wood or metal. For normal-weight concrete, wood is recommended. The bull float should be pushed ahead with the front (toe) of the float raised so that it will not dig into the concrete surface. The tool should be pulled back with the float blade flat on the surface to cut off bumps and fill holes.

Darbies are hand-operated wooden or metal tools, 3 to 8 feet long. For producing surfaces with close tolerances, low-handled wooden darbies are preferred. The darby should be held flat against the surface of the concrete and worked from right to left,

Bull floats are used to smooth large areas of concrete that are inaccessible to darbies.

or vice versa, with a sawing motion, cutting off bumps and filling depressions. When the surface is level, the darby should be tilted slightly and again moved from right to left, or vice versa, to fill any small holes left by the sawing motion.

The preceding operations should level, shape, and smooth the surface and work up a slight amount of cement paste. Do not overwork the concrete; overworking will result in a less durable surface.

Finishing Concrete

The finishing operations of edging, jointing, floating, troweling, and brooming must wait until all bleed water has left the surface and the concrete stiffens slightly. This waiting period, which is absolutely essential to obtain durable surfaces, varies with the ambient conditions—the wind, temperature, and relative humidity of the atmosphere—as well as the type and temperature of the concrete. On hot, dry, windy days, the waiting period is short; on cool, humid days, it can be several hours. With air-entrained concrete, there may be little waiting. Begin finishing when the water sheen is gone and the concrete can sustain foot pressure with

While pushing a bull float across a slab, tilt up the front edge to achieve proper compaction.

only about $1/4$-inch indentation. A water sheen might not be visible on air-entrained concrete.

A wooden darby is used after strikeoff to compact and level the fresh concrete.

Use a short, low-handled darby for more control when working a small area.

Edging

Edging is the first operation in finishing concrete. It produces a neat, rounded edge that prevents chipping or damage, especially when forms are removed. Edging also compacts and hardens the concrete surface next to the form where floats and trowels are less effective.

Edging tools are made of steel and bronze. Stainless steel edgers with a $1/2$-inch radius are recommended for walks, drives, and patios. Preliminary edging should be done with an edging tool that is wider and longer, approximately 6×10 inches.

Immediately after using the bull float or darby, cut the concrete away from the forms to a depth of 1 inch using a pointed mason's trowel or a margin trowel. Delay preliminary edging until the concrete has set sufficiently to hold the shape of the edger tool. The edger should be held flat on the concrete surface. The front of the edger should be tilted up slightly when moving the tool in the forward direction. When moving the tool backward over the edge, the rear of the tool should be tilted up slightly. Caution is necessary to prevent the edger from leaving too deep an impression, as these indentations may be difficult to remove in subsequent finishing operations. In some cases, edging is required after each finishing operation, with final edging done before or after brooming. Marks left by edging may be used for decorative purposes.

Jointing

During or immediately after edging, the slab is jointed or grooved. Jointing is a most important finishing operation, because proper jointing practices can eliminate unsightly random cracks.

Control joints, sometimes called contraction joints, can be made with a

Prior to edging, the concrete is cut away from the form with a pointed mason trowel.

Edges are formed with an edging tool. Use steady, even pressure as you pull or push the edger over the surface. Edging may be done after each finishing step.

hand tool, sawed, or formed by using divider strips. The tooled or sawed joint should extend into the slab one-fourth of the slab thickness. A cut of this depth provides a weakened section that induces cracking to occur beneath the joint where it is not visible.

Hand tools for jointing are called groovers or jointers. Like edgers, they are made of stainless steel and other metals and are available in various sizes and styles. The radius of a groover should be $1/4$ to $1/2$ inch. The bit (cutting edge) should be deep enough to cut the slab a minimum of one-fourth of the depth. Groovers with worn-out or shallow bits should not be used for making control joints, but may be used for decorative scoring of the surface.

It is good practice to mark the location of each joint with a string or chalk line on both side forms and on the concrete surface. A straight 1-inch board, wide enough to kneel on, should be used to guide the groover. The board should rest on the side forms. The groover should be held against the side of the board as it is moved across the slab.

To start the joint, the groover should be pushed into the concrete and moved while pressure is applied to the trailing edge of the tool. After the joint is cut, the tool should be turned around and pulled back over

the groove for a smooth finish. If the concrete has stiffened to the point where the groover will not penetrate easily to the proper depth, a hand axe or mason's trowel may be used to push through the concrete. The groover should then be used to finish the joint.

Instead of being hand-tooled, control joints are sometimes cut with a

On large concrete surfaces it may be more convenient to cut control joints with a power saw fitted with an abrasive or diamond blade.

Control joints can be made with a groover and should be cut about one-quarter of the way into the slab. Use a 1×6 as a straightedge to guide the tool.

On small jobs, control joints can be cut with a circular saw fitted with an abrasive blade. When using large or small power saws, be sure to wear eye protection.

concrete saw. Generally, this work must be done by a professional concrete sawing contractor. For small jobs, an electric handsaw equipped with a masonry cutting blade can be used. Saw joints as soon as the surface is firm enough not to be torn or damaged by the blade, normally 4 to 12 hours after the concrete hardens. A slight raveling of the sawed edges is permissible and indicates proper timing of the sawing operation. If sawing is delayed too long the concrete may crack before it is sawed or cracks may develop ahead of the saw.

Floating

Following edging and jointing, the concrete surface is floated. Floating has three purposes:

• To embed large aggregate just below the surface

• To remove any imperfections left in the surface by previous operations

• To compact the concrete and consolidate mortar at the surface in preparation for any further finishing operations that may be desired

Hand floats are made of metal, wood, plastic, or composition materials. Magnesium floats are light, strong, and slide easily over a concrete surface; they are recommended for most work, especially for air-entrained concrete. Wooden floats drag more, hence they require greater effort to use; but, wood produces a rougher texture, which may be preferred when good skid resistance is required and floating is used as the final finish. Power floats can be used to reduce finishing time on large slabs or for volume work. These machines, available at some tool rental stores, can be used for floating and troweling simply by changing blades.

The hand float should be held flat on the concrete surface and moved with a slight sawing motion in a sweeping arc to fill in holes, cut off lumps, and smooth ridges.

Floating produces a relatively even (but not smooth) texture. Because this texture has good skid resistance, floating is often used as a final finish and is recommended for exterior residential

For best results with air-entrained concrete, use a magnesium hand float to float the concrete surface before final troweling.

When working on knee boards, troweling is done at the same time as floating. Work one area at a time first with the float then finishing up with the trowel.

slabs. In such cases, it may be necessary to float the surface a second time after some hardening has taken place to impart the desired final texture to the concrete.

Marks left by edgers and groovers are removed during floating. Therefore, if the marks are desired for decorative purposes, the edger or groover can be rerun after final floating.

Troweling

After floating, the surface can be troweled. Troweling produces a smooth, hard, dense surface. But, for most outdoor slabs, a very smooth surface is neither necessary nor desirable. A troweled slab can be slippery when wet, and for many projects floating can provide the final finish. Troweling should never be done on a surface that has not been floated either by hand or power. Troweling after bull-floating or darbying is not sufficient.

On large slabs where you cannot reach the entire surface from outside the forms, you must work on knee boards. It is customary, when hand finishing, to float and immediately trowel an area before moving the knee boards. This operation should be delayed until the concrete has hardened enough that water and fine material are not brought to the surface. Too long a delay, of course, will result in a surface that is too hard to finish. The tendency in many cases, however, is to start when the concrete is too soft. Such premature finishing may cause scaling or dusting, as noted previously.

Hand trowels are made of high-quality steel in various sizes, and normally at least two sizes are used. For the first troweling, one of the larger tools ($4^3/4 \times 18$ inches, for example) is recommended. Shorter and narrower trowels are used for additional trowelings as the concrete sets and becomes

harder. A 3×12-inch trowel, known as a fanning trowel, is recommended for the final troweling. When one trowel is used for the entire operation, it should measure about 4×14 inches.

During the first troweling, the trowel blade should be held flat on the surface. If it is tilted, ripples are made that are difficult to remove later without tearing the surface. Move the trowel in a sweeping arc motion with each pass overlapping one-half of the previous pass. In this manner, each troweling will cover the surface twice. The first troweling may be sufficient to produce a surface free of defects, but additional trowelings may be used to increase smoothness and hardness.

There should be a lapse of time after the first troweling to permit the concrete to become harder. When only a slight indentation is made by pressing a hand against the surface, the second troweling should begin using a smaller trowel with the blade tilted slightly.

If the desired finish is not obtained with the second troweling, there should be a third troweling after another lapse of time. The final pass should make a ringing sound as the tilted blade moves over the hardening surface.

A smooth-troweled slab is easy to clean but can be slippery when wet. For better footing, it can be roughened slightly by brooming to produce a nonslip surface. (Brooming and other surface treatments are discussed in Chapter 5.)

Curing Concrete

Curing is one of the most important steps in concrete construction and, regretably, one of the most neglected. Proper curing greatly increases the strength and durability of concrete.

The hardening of concrete is brought about by chemical reaction between cement and water. This process, called *hydration*, continues only if water and a suitable temperature are available. When too much water is lost by evaporation from newly placed concrete, hydration stops. As the temperature drops to near freezing (32°F), hydration slows almost to a standstill. Under these conditions, concrete ceases to gain strength and other desirable properties. The purpose of curing is to maintain conditions under which concrete hardens, that is, to keep it moist and relatively warm.

Methods of Curing

Keeping concrete sufficiently moist for proper curing is done in a number of ways. For example, wet coverings, sprinkling, or ponding can be used to offset loss of moisture. Another approach is to cover the concrete surface with plastic sheeting, waterproof paper, or *curing compounds* to prevent moisture loss.

Supplying additional water is the most effective method. The most com-monly used wet covering is burlap, and it should be free of any substance that may be harmful to concrete or cause discoloration. Place the burlap as soon as the concrete is hard enough to withstand surface damage and sprinkle it periodically to keep the concrete surface continuously moist. A plastic sheet can be used to cover the wet burlap and keep it moist.

Water curing can be done with lawn sprinklers, nozzles, or soaking hoses; however, the concrete must be hard enough to resist the washing away of surface mortar. Also, the application of water must be continuous so that there is no chance of any partial drying during the curing period. Otherwise, alternate wetting and drying may cause cracks.

Curing methods intended to seal the surface are not quite as effective as methods that supply water, but they are widely used because of their convenience. Moisture barriers such as plastic sheeting and waterproof paper are popular, because they do not require periodic additions of water. However, they must be laid flat, thoroughly sealed at joints, and anchored carefully along edges.

Curing with plastic and paper materials may cause patchy discoloration, especially if the concrete contains calcium chloride and has been finished by hard steel troweling. This discoloration is experienced when the plastic or paper becomes wrinkled. Because it is difficult and time consuming on a project of significant size to smooth out the wrinkles that are apt to form, other means of curing should be used when uniform color is important.

In walls, forms provide satisfactory protection against loss of moisture if the top exposed concrete surfaces are kept wet. A soil-soaker hose is excellent for this. The forms should be left

Wet burlap is an efficient means for curing concrete as long as the burlap is kept constantly wet for about 7 days.

When curing with plastic sheeting, care must be taken to keep the plastic flat on the surface to prevent discoloration of the concrete.

on the concrete as long as practical. Wooden forms left in place should be kept moist by sprinkling, especially during hot, dry weather. If this is not done, they should be removed as soon as practical and another curing method started without delay.

Pigmented curing compounds provide the easiest and most convenient method of curing. These compounds are applied by spraying soon after the final finishing operation. The surface should be damp, but not wet with free water. Complete coverage is essential. A second coat, applied at right angles to the first coat is recommended.

Curing compounds are not recommended during late fall in regions where deicers are used to melt ice and snow. Use of curing compounds under these conditions may prevent proper

When using waterproof paper for curing, joints and edges should be weighted or taped to seal in moisture and slow down evaporation.

Curing concrete can also be done by spraying a curing compound on the concrete. A pigmented compound allows you to see where you've sprayed.

air drying of the concrete, which is necessary to enhance resistance to scaling caused by the use of deicers.

Curing should be started as soon as it is possible to do so without damaging the surface. It should continue for a period of 5 days in warm weather and 7 days in cool weather (50° to 70°F). The temperature of the concrete must not be allowed to fall below 50°F during the curing period. When moist curing has been completed, the concrete should not be subjected to forced or rapid drying. Drying out is a long, slow process: it takes concrete several months to dry out to ambient moisture conditions.

Protecting New Concrete

The surface of a newly completed slab should be protected against subsequent construction traffic. Equipment and workmen must not be allowed to damage the surface through neglect and carelessness. These rules should be followed.

• Prohibit foot traffic for 1 day.

• Prohibit light, rubber-tired vehicles for 7 days.

• Leave curing sheets in place as long as possible.

• Protect the surface with hard boards where heavy traffic is expected.

Cold-Weather Concreting

Ideally, walks, drives, patios, foundations, and steps should be built well in advance of cold weather. When placed during warm weather, there is plenty of time for concrete to develop the strength to resist freezing and thawing, and chemical deicers. However, a good job can also be done during cold weather by following the practices we will now discuss.

First of all, if you're making your own concrete, store all ingredients in a heated area if possible and use hot water for mixing. When ordering ready-mix, specify heated concrete so that the temperature of the mix does not fall below 50°F during placing, finishing, and curing.

When there is danger of freezing (particularly when average daily temperatures are below 40°F) concrete should be kept warm during curing. Insulating blankets or 12 to 24 inches of dry straw can be used. Straw should be covered with plastic sheeting to keep it dry and in place. The effectiveness of the protection can be checked by placing a thermometer on the concrete under the covering. Slab edges and corners are most vulnerable to freezing and should be carefully protected.

Use *high-early-strength* concrete to speed up the setting time and strength development. This can reduce the curing period from 7 to 3 days, though a minimum temperature of 50°F must be maintained in the concrete for those 3 days.

There are three ways to produce high-early-strength concrete.

• Use Type III or IIIA (high-early-strength) cement in the mix.

• Add an extra 100 to 200 pounds of cement to each cubic yard of concrete.

• Use a chemical accelerator, such as calcium chloride.

Because of problems that may occur with the use of chlorides, we recommend you follow either the first or second method, or combine the two.

Hot-Weather Concreting

In hot weather, precautions may be required to prevent rapid loss of surface moisture from the concrete. This can cause finishing difficulties and cracks (of the type known as plastic shrinkage cracking) in the fresh concrete during finishing or soon after. Here are some suggestions that will help prevent these problems.

• Dampen the subgrade and forms.

• Minimize finishing time by having sufficient manpower on hand.

• Erect sunshades and windbreaks.

• Use temporary coverings such as wet burlap or plastic sheeting during finishing and uncover only a small working area at a time.

Use of Deicers

Deicing chemicals used to melt ice and snow can cause surface scaling of non-air-entrained concrete. The extent of scaling depends on the amount of salt used and the frequency of use. Unfortunately, deicers can be applied accidentally in various ways, such as by drippings from the undersides of vehicles. Scaling may be much more severe in poorly drained areas because the deicer solution is retained on the concrete surface during repeated cycles of freezing and thawing.

Air entrainment (see table 2-2) has proved effective in preventing surface scaling and is recommended for all concretes exposed to freeze-thaw cycles or deicing chemicals. In addition, 4 weeks of air drying after the curing period will greatly increase concrete's resistance to deicers.

Ammonium nitrate and ammonium sulfate are sometimes sold as deicers. However, these particular materials react chemically with concrete and cause disintegration. Their use should be strictly avoided. Sodium chloride (common rock salt) is the safest deicer to use on concrete.

• Use light fog sprays above the surface to prevent evaporation.

• Start curing as soon as possible, using wet methods or white pigmented curing compounds.

• Place and finish during the cooler early morning or late evening hours.

CHAPTER 5

Decorative and Special Finishes

Many pleasing decorative finishes can be built into concrete during construction. Variations in the texture, pattern, and color of concrete surfaces are almost unlimited.

You can color concrete by adding pigments or by exposing colorful aggregates. Textured finishes can range from a smooth, polished appearance to the roughness of a gravel path. Geometric patterns can be scored or stamped into the concrete to resemble stone, brick, or tile paving. Other interesting patterns can be created by using wooden divider strips to form panels of various sizes and shapes: rectangular, square, or diamond. Special techniques are even available to render concrete sparkling or slip resistant.

What follows are brief descriptions of the various decorative and special finishes. As you read through the descriptions, keep in mind that different skills are necessary for each of the finishes. If you feel you need more information, consult a concrete contractor, your local ready-mix producer, or the Portland Cement Association.

Colored Concrete

Many decorative effects can be achieved by the use of colored concrete in and around the home for patios, floors, stepping-stones, walks, driveways, swimming-pool decks, and so on. There are four methods for producing colored concrete finishes: *one-course*, *two-course*, *dry-shake*, and *painting*.

The first three methods give the most satisfactory results. The one-course and two-course methods are similar. In both, the concrete mix is integrally colored by addition of a mineral oxide pigment especially prepared for use in concrete.

One-Course Method

In the one-course method, color pigment is mixed with the concrete to produce a uniform color throughout. The pigment can be a pure mineral oxide or a synthetic iron-oxide colorant especially prepared for use in concrete. Both natural and synthetic pigments are satisfactory if they are insoluble in water, free from soluble salts and acids, and fast to sunlight. Table 5-1 gives examples of some mineral pigments used in coloring concrete. These pigments are available from building materials suppliers.

Use the minimum amount of pigment necessary to produce the desired color, and never more than 10 percent by weight of the cement. Full-strength pigments will normally produce a good color when 7 pounds are mixed with 1 bag of cement; 1½ pounds per bag usually produces a pleasing pastel color. White portland cement will produce cleaner, brighter colors than normal gray portland cement.

To obtain a uniform color, all materials in the mix must be carefully proportioned by weight. Mixing time should be longer than normal to ensure uniformity. To prevent streaking, the dry cement and color compound should be thoroughly blended in the dry state before they are added to the mix. To prevent the contamination of a pigment mix by noncolored mixes, use separate mixers and thoroughly clean finishing tools when the work is complete.

When using the one-course method, uniform moistening of the subgrade is important for uniform color results. Soak the subgrade thoroughly the evening before you begin placing the concrete.

Two-Course Method

The two-course method uses a base course of conventional concrete and a topping of colored mortar (cement and sand). The base slab in the two-course method is placed in the usual manner as described in Chapter 4. To ensure a good bond, the topping course can be placed on the concrete as soon as it is firm enough to support a person's weight. The surface of the base course is left rough to help provide good bond for the topping. If the concrete must harden over a short period of time, the topping course can be bonded to the base concrete by either a cement-water grout or a reputable, proprietary bonding agent.

After the base stiffens slightly and the surface water disappears, a top course of ½ to 1 inch of colored mortar is placed. This top course is a medium-consistency cement-sand mixture to which colored mineral pigments are added, following the pigment manufacturer's specifications. The mix is floated and troweled in the usual manner. The thickness of the base slab

Table 5-1.	**Mineral Color Guide for Concrete Finishes**
Color Desired	**Materials to Use**
White	white portland cement, white sand
Brown	burnt umber or brown oxide of iron (yellow oxide of iron will modify color)
Buff	yellow ocher, yellow oxide of iron
Gray	normal portland cement
Green	chromium oxide (yellow oxide of iron will shade)
Pink	red oxide of iron (small amount)
Rose	red oxide of iron
Cream	yellow oxide of iron

must be adjusted to accommodate the topping course.

Dry-Shake Method

The dry-shake method consists of applying a dry color material that can be purchased from various manufacturers ready for use. Its basic ingredients are a pigment, portland cement, and specially graded silica sand or fine aggregate. After the concrete has been screeded and darbied or bull-floated, and excess moisture has evaporated, the surface should be floated.

Preliminary floating should be done before the dry-shake material is applied in order to bring up enough moisture for combining with the dry material. Floating also removes any ridges or depressions that might cause variations in color intensity. Immediately after floating, the dry material is shaken evenly by hand over the surface. If too much color is applied in one spot, nonuniformity in color and possibly surface peeling will result.

The first application of the colored dry-shake material should use about two-thirds of the total amount needed (in pounds per square foot as specified). In a few minutes this dry material will absorb some moisture

After dry-shake has been evenly distributed, it is floated into the surface with a magnesium float.

Dry-shake is distributed evenly by hand onto a floated slab. On the first application, use about two-thirds of the total amount of required material.

from the plastic concrete; it should then be thoroughly floated into the surface. Immediately, the balance of the dry-shake material should be distributed evenly over the surface, taking care that a uniform color is obtained. All tooled edges and joints should be "run" before and after the applications.

Troweling may follow the final floating if desired. After the first troweling, there should be a lapse of time—the length depending on such factors as temperature and humidity—to allow the concrete to increase its set. Then the concrete may be troweled a second time to improve the texture and produce a denser, harder surface.

For exterior surfaces, a second troweling is usually sufficient. Because smooth surfaces are slippery and hazardous when wet, draw a fine, soft-bristled broom over the surface to produce a roughened texture for better traction under foot. For interior surfaces a third hard troweling may be needed.

Painting Concrete

Normally, paints are used only when it's necessary to color existing concrete. Paints on areas subject to heavy traffic and especially on outdoor paving wear away eventually and must be renewed if a good appearance is to be maintained.

When painting concrete, surface preparation is important. Fresh concrete should be well cured for 28 days to 6 months depending on the type of paint to be applied, because moisture in the concrete can cause some paints to blister and peel. Also, be sure that the surface is free of dirt, grease, oil, laitance, and *efflorescence*. Dirt is removed with water and a brush; grease and oil cleans off with trisodium phosphate (TSP) and a thorough rinse; laitance (a fine powderlike material sometimes found on hardened concrete) must be removed with steel brushes or acid etching; and efflorescence requires washing with a 5- to 10-percent solution of muriatic acid. Be sure to handle cleaning chemicals carefully and follow all of the manufacturer's instructions.

Concrete cast against forms is sometimes so smooth that it makes adhesion of coatings difficult to obtain. Such surfaces should be pressure washed or ground with silicon carbide stones to provide a slightly roughened surface. Whenever concrete is to be painted, forms should be coated with a release agent that will remain on the form when it is stripped.

If you plan on painting floors, you should not use any curing compounds on them because these compounds form a film that interferes with the bond between paint and concrete.

Two types of coating performance are necessary for painting concrete: one for below grade where water may be present, the other for above grade where wind and rain have an effect on the paint. Concrete paint should be either of the breathing type, permeable to water vapor but impermeable to liquid water, or impermeable to all water. Normally, exterior surfaces of concrete should be able to breathe to permit water vapor to travel through walls and floors preventing both the blistering of paint due to excessive vapor pressure and the spalling of masonry due to freezing of entrapped moisture. Interior surfaces, on the other hand, should have a barrier to prevent water vapor inside the building from entering the wall.

Types of paint available for concrete have bases such as portland cement, oil, latex, rubber, and epoxy resin. Because paint technology is constantly evolving, consult a reputable painting contractor, a paint store salesperson, or the Portland Cement Association for the type of paint best suited for your application.

Exposed Aggregate

Exposed aggregate is one of the most popular decorative finishes. It offers unlimited color selection and a wide range of texture. Exposed-aggregate finishes are not only attractive, but they can be rugged, slip resistant, and highly immune to wear and weather. They are ideal for sidewalks, driveways, patios, pool decks, and other flatwork applications.

Aggregates for an exposed surface should fall into narrow ranges (see table 5-2). The proper depth of exposure for an exposed-aggregate surface should not exceed one-third of the average

Table 5-2. **Aggregate Sizes for Exposed-Aggregate Slabs**
$1/4''$ to $1/2''$
$3/8''$ to $5/8''$
$1/2''$ to $3/4''$
$5/8''$ to $7/8''$
$3/4''$ to $1''$
$1''$ to $1 1/2''$

diameter of the aggregate and not more than one-half of the diameter of the smallest aggregate.

The Seeding Method

There are a number of ways to obtain exposed-aggregate finishes, but the seeding method is one of the most practical and commonly used techniques. In this method, special aggregates are seeded into the top of the slab rather than mixed throughout the concrete. Because of the extra working time required with exposing aggregates, you may want to use a surface retarder for better control of the exposing operations; however, this step is not essential.

Begin by setting forms for a conventional concrete slab. If the slab is large, you may need to add construction joints because the seeding method takes about three times longer than normal finishing and you'll need to work on smaller sections at a time. When building the forms for a seeded slab, set the height $^1/_8$ to $^7/_{16}$ inch lower than the desired final finished elevation of the slab to allow for the added

For an exposed-aggregate finish, strike off, darby or bull float slabs in the usual manner. Then spread selected aggregate uniformly by shovel and hand so that the entire surface is completely covered with a single layer of stone.

thickness of the seeding aggregate (see table 5-3).

Use concrete with a slump of 3 to 4 inches and a maximum $^3/_4$-inch size of coarse aggregate to ease embedment of the seeding aggregate. The aggregate to be exposed must be carefully selected to ensure that it does not contain such substances as iron oxides and iron pyrite because these stain concrete surfaces. For best results, select rounded river gravel and avoid crushed stone.

After the slab is struck off and darbied or bull-floated in the usual manner, spread the aggregate over the slab by hand or shovel so that the surface is completely and uniformly covered with one layer of aggregate.

The seeded aggregate is embedded in the concrete by tapping it with a wooden hand float, darby, or straightedge. Final embedment is done by using a magnesium float until all of the aggregate is entirely embedded and mortar completely surrounds and slightly covers all particles. The surface appearance after all the aggregate is embedded

Table 5-3.	**Strikeoff Allowance and Aggregate Size for Seeding Method**
Strikeoff Allowance	**Aggregate Size**
$^1/_8''$	$^3/_8''$ to $^5/_8''$
$^3/_{16}''$	$^1/_2''$ to $^3/_4''$
$^1/_4''$	$^3/_4''$ to $1''$
$^5/_{16}''$	$1''$ to $1^1/_2''$
$^7/_{16}''$	$1^1/_4''$ to $2''$

will be similar to that of a normal slab after floating, with all voids and imperfections removed.

Ideally, all the embedded aggregate should be worked down until it is covered with a layer of mortar about $1/16$ inch thick. Care must be taken not to embed the aggregate too deeply and to ensure that the finished surface remains flat and is not deformed.

None of the seeding aggregate should become intermixed with the aggregates of the base concrete during the embedment. If this happens, the coarse aggregate in the base concrete will show up on the finished surface. Any need for additional mortar to embed the aggregates is generally due to an improper mix design or delaying the seeding and embedment operation.

The timing of the start of the aggregate exposure is critical and is usually based on previous experience. If you have no previous experience, you may want to start with a very small or inconspicuous area just to get the feel of this process. In general, exposing aggregate should be delayed until the slab will bear the weight of a person on knee boards with no indentation. At this time, the slab is lightly brushed with a stiff nylon-bristled broom to remove excess mortar. If this brushing dislodges any aggregate, delay the operation until the resumption of brushing does not cause this to happen.

After an initial brushing, brush the surface again while simultaneously flushing away loosened mortar with a fine water spray. As the slab continues to set, washing and brushing can proceed at a more vigorous pace. (Soft- and hard-bristled brooms and special exposed-aggregate brooms with attached water jets are available for this job.) The washing and brushing should con-

Embed the aggregate initially by tapping with a wooden hand float, a darby, or a straightedge.

For final embedding, use a bull float or hand float until the appearance of the surface is similar to that of a normal slab after floating. The aggregate should be completely surrounded, but only slightly covered with mortar.

tinue until exposure is uniform and at the proper depth, the flush water runs clear, and there is no noticeable cement film on the aggregate.

Timing of the start of the aggregate exposure operation is critical. In general, wait until the slab can bear the weight of a person on knee boards with no indentation. Then brush the slab lightly with a stiff nylon-bristled broom to remove excess mortar.

Next fine-spray with water along with brushing. Special exposed-aggregate brooms with water jets are available. If aggregate is dislodged, delay the operation, otherwise continue washing and brushing until flush water runs clear and there is no noticeable cement film left on the aggregate.

The Topping Course Method

In this method, a thin topping course of concrete containing special aggregates is placed over a base slab of conventional concrete. The topping normally is 1 to 2 inches thick depending on the aggregate size. The base slab is struck off below the top of the forms to leave room for the topping course. The surface of the base course should have a rough-broomed finish and be firm enough to support a person's weight before the topping is placed.

This method of construction is suitable when smaller aggregate gradations are used such as $1/4$ to $1/2$ inch or $3/8$ to $5/8$ inch. The topping uses a specially designed mixture of selected gap-graded aggregates and masonry sand instead of normal concrete sand. (In a gap-graded mix, the coarse aggregate is all the same size or limited to a few sizes rather than a range of sizes as in conventional concrete.) Use of a surface retarder is advisable with these smaller aggregate sizes. The same brushing and washing procedure of aggregate exposure is utilized in this type of construction as with the seeding method.

Terrazzo construction uses this topping technique. Terrazzo toppings for outdoor work are $1/2$ inch thick and contain decorative aggregates such as marble, quartz, or granite chips. The colorful aggregates are exposed by brushing and washing with water. Brass or plastic-topped divider strips set in a bed of mortar are used to eliminate random cracking. They also permit the use of different colored terrazzo mixtures in a wide variety of patterns. This type of terrazzo is called rustic or washed terrazzo.

NOTE: Because terrazzo requires even more skill than normal topping, it should only be undertaken by a qualified terrazzo contractor.

Exposing Aggregates in Conventional Concrete

An alternate method consists of exposing the coarse aggregate in conventional concrete. For this method to work, the mix must contain a high proportion of coarse aggregate. Typically, the ratio for the total aggregate content is 70 percent coarse to 30 percent fine. The coarse aggregate used should be uniform in size (gap-graded), bright in color, closely packed, and properly distributed. The concrete slump must be low (1 to 3 inches) so that the coarse aggregate remains near the surface.

In placing, striking off, bull-floating, or darbying, the usual procedures are followed. Care should be taken not to overdo floating, as this may depress the coarse aggregate too deeply. The aggregate is ready for exposing when the water sheen disappears, the surface can support a person's weight without indentation, and the aggregate is not overexposed or dislodged by washing and brushing. When using smaller aggregate sizes, it is helpful to use a surface retarder to delay the time of set of the surface to allow the subsurface concrete to harden. This procedure will help prevent dislodgment of the small-size aggregates.

As soon as the water sheen has disappeared, the retarder is sprayed over the surface with an ordinary, low-pressure garden sprayer. Normally, 1 gallon of retarder will cover from 100 to 150 square feet of concrete. The surface should then be covered with plastic sheeting to continue curing. If the retarded area is to remain undisturbed overnight, check a small area (in the morning) to determine if the concrete has hardened sufficiently.

Cleaning Exposed-Aggregate Surfaces

Various techniques and materials to clean exposed-aggregate concrete include washing it with plain water or hot water containing detergents or other commercial cleaners, steam, light abrasive blasting, or application of dilute hydrochloric (muriatic) acid.

The use of an acid-wash surface treatment is not a necessity, but it often helps to brighten the appearance of an exposed-aggregate surface, especially those made with darker aggregates. For best results, acid washing should be delayed a minimum of 2 weeks after the concrete has been placed; a longer delay is better. Caution should be exercised when using acid on some aggregates such as limestones, dolomites, and marbles, which may discolor and dissolve in the muriatic acid. The surfaces to be cleaned with a weak acid solution should be thoroughly saturated with water, and any excess water should be removed before application of the acid.

WARNING: Wear protective clothing and protect adjacent areas and materials from the acid. Residue from acid washing should be flushed with clear water and drained away from areas that might be damaged.

An efficient acid-etching solution can be made with 1 part muriatic acid and 2 to 4 parts of fresh water. If a chloride-free acid-etching solution is required, 85 percent phosphoric acid diluted with 2 to 3 parts of fresh water may be used. Stronger acid-etching solutions may be used if the etching action is insufficient, but care should be taken not to etch the concrete so deeply that the aggregate is not securely embedded. The acid is brushed on the surface and thoroughly rinsed off. The rinsing operation should begin when the bubbling action of the acid begins to subside. The best procedure is to work with a helper so one person can brush while the other one rinses.

Sealing Exposed-Aggregate Surfaces

Clear coatings bring out the true color of the aggregate and help keep exposed surfaces from discoloring with use. However, care is needed in the selection of a coating material. Some coating materials, such as linseed oil, may darken the matrix. Other coatings may oxidize from exposure to sunlight and become a dirty yellow, or possibly brown. The best coating materials consist of *acrylic resins*.

Textured Finishes

Decorative textures can be produced on a concrete slab with little effort and expense by using floats, trowels, and brooms. More elaborate textures are possible with special techniques using a mortar dash coat or rock salt.

Float and Trowel Textures

A swirl finish lends visual interest as well as surer footing. To produce this texture, the concrete is struck off, bull-floated (or darbied), and hand floated in the usual manner except that the float should be worked flat on the surface in a swirling motion using pressure. Different patterns are made by using a series of uniform arcs or twists. Coarse textures are produced with wooden floats; medium textures with aluminum, magnesium, or can-

You can produce a swirl finish on concrete with a hand float or trowel.

vas resin floats. A smooth-textured swirl is obtained with a steel trowel.

To produce a swirl trowel texture, also known as a *sweat finish*, the concrete is finished as above and then troweled in the usual manner. Sometime after the first troweling, lay a trowel flat on the surface and work it around in a swirling motion while applying pressure. It's best to use an older, "broken-in" trowel for this procedure because its blade will have a slight curvature. This will allow the trowel to be worked flat without the edges digging into the concrete.

Timing is important when producing a swirl trowel finish. After the first troweling, there should be a lapse of time—just how much depends on such factors as temperature and humidity —to allow the surface to become harder. Don't wait too long, because the swirl should be applied while it is still possible to work a small amount of fine mortar to the surface. This material creates a drag on the trowel, leaving the surface with a fine-textured, matte-like finish. After creating this texture, make sure the concrete has set sufficiently that the texture will not be marred before beginning the curing process.

Broomed Finishes

Broomed finishes are attractive, nonslip textures secured by pulling damp brooms across freshly floated or troweled surfaces. Coarse textures suitable for steep slopes or heavy traffic are produced by stiff-bristled brooms on newly floated concrete. Medium to fine textures are obtained when using soft-bristled brooms on floated or steel-troweled surfaces. Best results are obtained when using a broom that is specifically made for texturing concrete.

A broomed texture can be applied

in many ways: straight lines, curved lines, or wavy lines. Driveways and sidewalks are usually broomed at right angles to the direction of traffic. Each pass of the broom can be contiguous to the previous one, or an area of unbroomed concrete can be left between passes. To create a checkerboard effect, the slab is divided into square panels by means of control joints and each panel is broomed at a 90-degree angle to the brooming of immediately adjacent panels.

Travertine Finishes

A *travertine* finish, sometimes called keystone, requires a more elaborate procedure. After the concrete slab has been struck off, bull-floated or darbied, and edged in the usual manner, the slab is broomed with a stiff-bristled broom to ensure bond when the finish (mortar) coat is applied.

The *finish coat* is made by mixing 1 bag of white portland cement and 2 cubic feet of sand with about 1/4 pound of color pigment. Yellow (yellow ocher) is generally used to tint the mortar coat, but any mineral oxide color may be used. Care must be taken to keep

This broom is especially made for fine-textured broomed finishes. It is 18 inches wide with 2¹/₄-inch plastic bristles set into a ⁷/₈-inch-thick wooden holder.

By using a broom, you can create a wavy texture as shown here.

the proportions exactly the same for all batches. Add enough water to make a soupy mixture having the consistency of thick paint.

This mortar is placed in pails and thrown vigorously on the slab with a dash brush to make an uneven surface with ridges and depressions. The ridges should be about 1/4 to 1/2 inch high. The mortar is allowed to set enough to permit a cement mason to work on the surface with knee boards.

For homes in warmer climates, a travertine finish can add beauty and interest to concrete walks and patios.

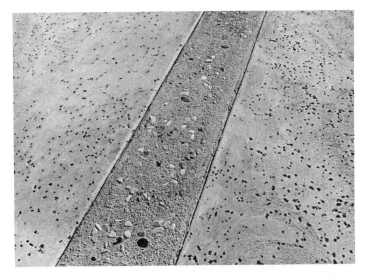

Rock salt texture is created by embedding and floating selected particles of rock salt into the fresh concrete and curing the slab using a building plastic or waterproof paper.

The slab is then steel troweled to flatten the ridges and spread the mortar. This leaves the surface smooth at its highest points, but rough in the voids and low areas. When the mortar is made with white cement and yellow pigment, the effect produced resembles the appearance of travertine marble. By varying the amount of mortar thrown on the slab and the extent of troweling, many interesting texture variations can be produced. For added interest, the slab may then be *scored* into random geometric designs before curing.

Rock Salt Texture

A rock salt texture is frequently used as an economical, decorative surface finish for concrete slabs. This texture is produced by scattering rock salt over the surface after floating, troweling, or brooming. The salt grains are rolled or pressed into the surface with only the tops of the grains left exposed. After 5 days of curing under plastic film or waterproof paper, the surface is washed and brushed, dislodging and dissolving the salt grains and leaving pits or holes. The surface between the holes may be left slightly rough to give better traction in outdoor applications.

The salt crystals used to create this finish are ordinary water-softener salt crystals and should all be between $1/8$ and $3/8$ inch in size. Using coarser rock salt yields a greater percentage of larger holes in the finished surface texture; however, holes larger than $1/4$ inch are not desirable for foot traffic. The salt grains normally are distributed at a rate of about 5 pounds per 100 square feet of slab surface, but a range of 3 to 12 pounds per 100 square feet may be used to create a light to heavy pattern. The heavier concentrations produce textures that resemble travertine.

Hand tools or pipes may be used to press or roll the salt grains into the surface, but a 90-pound floor tile roller works best. The grains must be spread and worked into the surface before the concrete becomes too hard.

Curing compounds may be used in lieu of sheet curing materials, but the salt-dissolving operation will be more difficult. A water cure is not satisfactory for use with the rock salt procedure, and plastic sheeting may cause discoloration if it is not kept flat and wrinkle free on the concrete surface.

CAUTION: Neither the rock salt nor the travertine finish is recommended for use in areas subject to freezing weather. Water trapped in the recesses of these finishes tends, when frozen, to spall the surface.

Geometric Patterns

A wide variety of geometric designs can be stamped, scored, or sawed into a concrete slab to enhance the beauty of walks, drives, and patios. Random flagstone or ashlar patterns may be produced by embedding 1-inch-wide strips of 15-pound roofing felt in concrete. After floating, the strips (precut to the pattern desired) are laid on the surface, patted flush, and floated

over. At this time, a color dry-shake can be applied or the slab can be finished in its natural color. The strips are carefully removed before the slab is cured.

An alternate method of producing these patterns makes use of an 18-inch-long piece of $^1/_2$- or $^3/_4$-inch copper pipe bent into a flat S-shape. Scoring must be done while the concrete is still plastic because coarse aggregate must be pushed aside. The best time is soon after darbying (or bull-floating). A second scoring to smooth the joints is done after hand floating.

Pattern Stamped Finishes

Stone, brick, tile, and other patterns can be deeply cut into partially set concrete with special stamping tools. The concrete may be colored integrally or by the dry-shake method and the joints filled with plain or colored mortar to create any number of striking effects.

To receive a stamped pattern, concrete should contain small coarse aggregate such as pea gravel ($^3/_8$-inch top size). Finishing follows the normal procedures; however, do not trowel the surface more than once. After the sur-

A random flagstone pattern on a sidewalk or patio is created using a bent piece of $^1/_2$-inch copper pipe or by embedding 1-inch-wide strips of 15-pound felt. When using the felt, embed it before final floating and remove it after troweling.

Stamped patterns can be created with specially made stamping pads just after the concrete surface is floated or troweled. Pads come in a variety of shapes and at least two are required. Tamp the pads to ensure proper indentation.

face is floated or troweled to the desired texture, platform stamping pads are used. One pad is placed next to the other so that the pattern is accurately aligned; at least two pads are required. The user simply steps from one pad to the next, stamping the design to a depth of about 1 inch. Brush a form release agent on the pads to keep them free of mortar.

In addition to your weight, a hand tamper is sometimes used to ensure adequate indentation of the stamping pad. After stamping, a tool similar to a brick mason's jointer is used to dress edges and cause some artificial imperfections to give a natural look. A small hand stamp is used to complete the pattern next to the slab edges. Light tapping with a hammer may be required. This operation can be eliminated or greatly minimized by choosing slab sizes that are equal to even multiples of the platform stamping tool dimensions. Timing is critical because all stamping must be completed before the concrete hardens completely.

Other interesting relief patterns can be stamped into the surface of concrete slabs to resemble a wood-grained boardwalk, a basket weave, or seashells. Check with your local tool rental service about the availability of these stamps. If they are unavailable for rent, you may buy them through a mason's tool supply, or hire a mason or concrete contractor to perform this particular operation.

Scoring and Sawing

Straight-line patterns may be scored into concrete slabs with an ordinary cement mason's grooving tool when the concrete is fresh. Patterns may also be cut into the surface with a power saw after the concrete hardens.

When scoring with a groover, the use of a straightedge such as a 1-inch-

A small hand stamp is used to complete the pattern next to the pavement edges. Light tapping with a hammer may be required.

thick board that is at least 6 inches wide is recommended as a guide, following the same process for cutting control joints. The bit cutting edge should be deep enough to cut the slab to a depth of about $1/4$ inch for decorative work. However, if the groove is also to serve as a control joint, a groover with a bit deep enough to cut the slab a minimum of one-fourth of the depth of slab is mandatory.

For small jobs, an electric handsaw equipped with a masonry cutting blade can be used to cut decorative grooves. Sawing should be done as soon as the surface is firm enough not to be torn or damaged by the blade, which is normally 12 hours after the concrete hardens.

Nonslip and Sparkling Finishes

Surfaces that are frequently wet or that would be dangerous if slippery can be given special nonslip finishes. These finishes are achieved most commonly by hand tools such as floats, trowels, or brooms, or with dry-shake applications of abrasive grains. The latter method provides a long-lasting, nonslip surface suitable for areas with heavy foot traffic.

The two most widely used abrasive grains are silicon carbide and aluminum oxide. Silicon carbide grains are sparkling black in color and are also used to make "sparkling concrete." The sparkle is especially effective under artificial light. Aluminum oxide may be gray, brown, or white and is used where the sparkle of silicon carbide is not required. Application of the abrasive grains follows essentially the same procedure as for a color dry-shake. The grains should be spread uniformly over the surface in a range of from $\frac{1}{4}$ to $\frac{1}{2}$ pound per square foot and lightly troweled. The manufacturer's directions should be followed.

Combination Finishes

In using concrete decoratively, striking effects can be obtained by combining colors, textures, and patterns. For example, alternate areas of exposed aggregate can be eye catching when combined with plain, colored, or textured concrete. Ribbons and borders of concrete masonry or brick add a distinctive touch when combined with exposed aggregate. Also, light-colored strips of exposed aggregate may divide areas of dark-colored concrete or vice versa. Scored and stamped designs are enhanced when combined with integral or dry-shake color.

What we have described in this chapter are just a few of the possibilities available for applying special textures and finishes to concrete surfaces. With a little imagination, concrete driveways, sidewalks, and patios can be tailored to fit the mood and style of any architecture or landscape.

CHAPTER 6

Concrete Defects

When you consider the vast amount of concrete used in construction, the number of complaints that arise are amazingly few. Still, a variety of potential problems do exist and it is good to be able to recognize them and understand their causes. It's fairly easy to predetermine the properties desired in the concrete used for a particular purpose, but great care is required throughout the entire process of mixing, placing, finishing, and curing concrete to ensure that the hardened concrete actually ends up with the desired properties.

If a blemish appears on the surface of a concrete slab, most likely it will fall into one of the following categories: dusting, scaling, popouts, crazing, cracking, discoloration, or spalling. These defects, caused by factors we will now explain, can be prevented by adhering to proper construction methods.

Some repair tips are included in this discussion, but its primary purpose is to describe potential problems, their causes, and how to avoid them.

More detailed information on the maintenance and repair of concrete is provided in Chapter 7.

Dusting

Dusting, the development of a fine, powdery material that easily rubs off the surface of hardened concrete, can occur either indoors or outdoors; but, it's more likely to be a problem when it occurs indoors. Dusting is the result of a thin, weak layer, called *laitance*, composed of water, cement, and fine particles.

Fresh concrete is a fairly cohesive mass with aggregates, cement, and water uniformly distributed throughout. A certain amount of time must elapse before the cement and water react sufficiently for the concrete to harden. During this period, the cement and aggregate particles are partly suspended in the water. Because the cement and aggregates are heavier than the water, they tend to sink. As they move downward the displaced water moves upward and appears at the sur-

Dusting is evident by the fine powder that can be easily rubbed off the surface.

face as bleed water, with the result that there is more water near and at the surface than in the lower portion of the concrete. This means that the weakest, most permeable, and least wear-resistant concrete (the laitance) is at the top surface, exactly where we need the strongest, most impenetrable, and most wear-resistant concrete.

Floating and troweling concrete that has this bleed water on it mixes the excess water back into the surface, further weakening its strength and wear resistance. This is one of the principal causes of dusting.

Broadly speaking, dusting may be caused or aggravated by any of the following practices or situations:

• Floating and troweling bleed water back into the concrete surface

• Applying water during finishing

• Using cement as a dry-shake to speed up drying

• Using dirty aggregates

• Using a mix with a low cement content

• Using a mix that is too wet

• Curing improperly (especially, allowing the surface to dry rapidly)

• Freezing of the surface

One way to correct a dusting surface is to grind off the thin layer of laitance to expose the solid concrete underneath. Another method is to apply a chemical surface *hardener.* A surface hardener will not convert a basically bad concrete slab into a good one, but it will improve wearability and reduce dusting of the surface. The major ingredient in many floor-surface hardeners is sodium silicate (water glass) or a metallic fluosilicate (such as magnesium and zinc fluosilicates). The treatment is usually applied in two or three coats, and the surface is allowed to dry between each coat.

Scaling and Mortar Flaking

Scaling is the loss of surface mortar and mortar surrounding the aggregate particles. The aggregate is usually clearly exposed and often stands out from the concrete. Scaling is primarily a physical action caused by hydraulic pressure from water freezing within the concrete and not usually caused by chemical corrosive action. When pressure exceeds the tensile strength of concrete, scaling can result if entrained air voids are not present to act as internal pressure relief valves.

Scaling is a scabrous condition where the surface mortar has peeled away, usually exposing the coarse aggregate.

The presence of a deicer solution in water-soaked concrete during freezing causes an additional buildup of internal pressure. Deicers such as sodium chloride (rock salt), urea, and weak solutions of calcium chloride do not chemically attack concrete; however, deicers containing ammonium sulfate or ammonium nitrate will rapidly disintegrate concrete and should not be used. Several deicers, particularly those containing chloride, can also cause corrosion of embedded steel.

Mortar flaking over coarse aggregate particles is a form of scaling that somewhat resembles a surface with popouts. However, mortar flaking usually does not result in freshly fractured aggregate particles, and there are few, if any, cone-shaped holes such as those found in popouts. Aggregate particles with flat surfaces are more susceptible to this type of defect than round particles. Mortar flaking occasionally precedes more widespread surface scaling, but its presence is not necessarily an indication of more extensive scaling.

Mortar flaking over coarse aggregate particles is caused essentially by the same actions that cause regular scaling and often results from placing concrete on hot, windy days. Excessive and early drying out of the surface mortar can alone aggravate scaling; however, the moisture loss is accentuated over aggregate particles because bleed water beneath the aggregate cannot readily migrate to the surface to replenish the evaporated water. This combination of bleed-water blockage, high evaporation, and lack of moisture necessary for cement hydration results in a dry mortar layer of low strength, poor durability, high shrinkage, and poor bond with the aggregate. If this thin, weakened mortar layer becomes saturated with moisture, then freezes, it will break away from the aggregate. Poor finishing practices can also aggravate mortar flaking.

Preventing Scaling and Flaking

Experience and testing have proved that when air-entrained concrete is properly proportioned, placed, finished, and cured it is highly resistant to surface scaling and mortar flaking due to freezing and thawing and the application of deicer chemicals. The main problem with scaling and flaking is that they may not occur until several years after you've completed the concrete work. Here are some tips to help you prevent these problems from ever developing.

• Use a proper concrete mix with durable and well-graded aggregate, at least 6 bags of cement per cubic yard, 5 to $7\frac{1}{2}$ percent air content (see table 2-2), and a 3- to 5-inch slump. Wet, sloppy, low-strength concrete with low cement content and without entrained air voids is prone to scale.

• Properly slope the concrete to drain water away from the slab. Saturated concrete is much more susceptible to deterioration than drier concrete.

• Use proper finishing practices. Concrete that is prematurely floated or

This is an example of mortar flaking over coarse aggregate.

troweled while bleed water is on the surface tends to scale when subjected to saturated freezing. In addition, finishing concrete before the bleed water comes to the surface can entrap water under the finished surface, forming a weakened zone or void, which can result in scaling or surface delamination.

• Cure promptly with wet burlap (kept continuously wet), curing paper, or plastic sheets until the concrete is 5 to 7 days old.

• After curing, allow the concrete 30 days to air dry before it is exposed to freezing and thawing or deicers.

If scaling or mortar flaking should develop, or if you suspect the concrete is poor in quality, a breathable surface treatment may be applied to help protect it against further freeze-thaw damage. These treatments are made with linseed oil, silane, siloxane, methacrylate-based compounds, or other materials.

Oil treatment normally consists of two applications of equal parts of commercial boiled linseed oil and a solvent such as turpentine, naphtha, or mineral spirits. Recommended coverage is 40 to 50 square yards per gallon for the first application and about 70 square yards per gallon for the second application. To assure proper penetration and hasten drying, apply this mixture when the temperature is 50°F or above. And because oil treatment produces a slippery surface until absorbed, keep traffic off the concrete until sufficient drying has taken place.

Impenetrable materials, such as most epoxies, should not be used on slabs on grade or other concrete where moisture can freeze under the coating. The freezing water can cause delamination under the impenetrable coating; therefore, a breathable surface treatment should be used.

Thin-bonded overlays or surface-grinding methods can usually remedy a scaled surface if sound, air-entrained concrete is present below the scaled surface.

Popouts

A *popout* is a cone-shaped fragment that breaks out of the surface of the concrete leaving a hole that generally varies in size from 1/4 to 2 inches. Usually a fractured aggregate particle will be found at the bottom of the hole with part of the aggregate still adhering to the point of the popout cone.

The cause of a popout is usually a piece of porous aggregate having a high rate of *absorption* and located near the concrete surface. As the porous aggregate absorbs moisture or freezing occurs under moist conditions, its swelling creates internal pressures sufficient to rupture the concrete surface. Popouts can also be caused by chemically reactive aggregates in the concrete.

Most popouts appear within the first year, occurring shortly after placement due to the absorption of water from the freshly mixed concrete, or after a season or year of high humidity or rainfall. Popouts are considered a cosmetic detraction and generally do not affect the service life of the concrete.

A popout is a small fragment of concrete surface that breaks away due to internal pressure, leaving a shallow, typically conical, depression.

Popout Prevention and Repair

The following steps can be taken to minimize or eliminate popouts.

• Use concrete with the lowest possible water content and slump for the application.

• Use air-entrained concrete.

• Use proper aggregates.

• Cover the surface with plastic sheets or wet burlap after screeding and bull-floating during hot, dry, and windy weather to reduce the rate of evaporation.

• Do not finish concrete with bleed water on the surface.

• Avoid hard steel troweling where not needed, such as on exterior slabs.

• Avoid use of *vapor barriers;* but if one is required, cover it with 2 to 3 inches of damp sand before the concrete is placed.

• Use wet-curing methods such as continuous sprinkling with water or covering with wet burlap soon after final finishing.

• Wet cure for a minimum of 7 days and flush curing water from the surface before final drying.

• Avoid impervious floor coverings or membranes as they can aggravate popout development.

• Slope slab surfaces to drain water properly.

Surfaces with popouts can be repaired. A small patch can be made by drilling out the spalled particle and filling the void with a dry-pack mortar, epoxy mortar, or other appropriate patch material. If the popouts are too numerous to patch individually, a thin-bodied overlay or surface grinding may be used to restore serviceability.

Crazing

Crazing is a network pattern of fine cracks that do not penetrate much

Crazing is a network of fine, superficial surface cracks.

below the surface and are caused by minor surface shrinkage. *Craze cracks* are very fine and barely visible except when the concrete is drying after the surface has been wet. The cracks encompass small concrete areas less than 2 inches on a side, forming a chicken-wire pattern. The term *map cracking* is often used to refer to cracks that are similar to craze cracks only more visible and surrounding larger areas of concrete. Although craze cracks are unsightly and can collect dirt, crazing is not structurally serious and does not ordinarily indicate the start of future deterioration.

When concrete is just beginning to gain strength, the climatic conditions, particularly the relative humidity during the drying period in a wetting and drying cycle, are an important cause of crazing. Low humidity, high air temperature, hot sun, or drying wind, either separately or in any combination, can cause rapid surface drying that encourages crazing. A surface into which dry cement has been cast to hasten drying and finishing will be more subject to crazing. The conditions that contribute to dusting, as discussed earlier, also increase the tendency to craze.

To prevent crazing, curing procedures should begin early, within minutes after final troweling, when weather conditions are adverse. When the temperature is high and the sun is out, some method of curing with water should be used because this will stop rapid drying and lower the surface temperature. The concrete should be protected against rapid changes in temperature and moisture wherever feasible.

Cracking

Unexpected *cracking* of concrete is a frequent cause of complaints. Cracking can be the result of one or a combination of factors such as drying shrinkage, thermal contraction, restraint (external or internal) to shortening, subgrade settlement, and applied loads. Cracks occurring before the concrete hardens are usually the result of settlement within the concrete mass or shrinkage of the surface (plastic-shrinkage cracks) caused by the rapid loss of water while the concrete is still plastic. Cracking can be significantly reduced when the causes are recognized and preventive steps taken. For example, joints provided in the design and installed during construction force cracks to occur in places where they are inconspicuous.

Settlement cracks may develop over embedded items, such as reinforcing steel, or adjacent to forms or hardened concrete as the concrete settles or subsides. Settlement cracking results from insufficient consolidation (vibration), high slump (overly wet concrete), or lack of adequate cover over embedded items.

Plastic-shrinkage cracks are relatively short, shallow cracks that develop before final finishing because of rapid evaporation. Usually, hot, dry, windy weather is the cause of the rapid evaporation. But, the problem can occur in cool weather when there is concentrated artificial heat present along with high winds and low humid-

Plastic-shrinkage cracks are caused by rapid loss of mix water while the concrete is still plastic.

ity. When surface moisture evaporates faster than it can be replaced by rising bleed water, the surface shrinks more than the interior concrete. As the interior concrete restrains the surface shrinkage, stresses develop that exceed the concrete's tensile strength. This leads to cracks that often penetrate to the center of a slab.

Cracks that occur after hardening are usually the result of drying shrinkage, thermal contraction, or subgrade settlement. While it hardens, concrete shrinks about $1/16$ inch every 10 feet. This amount of shrinkage creates a tensile stress about three times greater than the ultimate tensile strength of concrete. If the concrete is restrained, it will crack. Experience has shown that control joints (induced cracks) should be spaced at about 10-foot intervals in each direction in 4-inch-thick unreinforced concrete slabs-on-grade (see table 3-2).

The major factor influencing the drying-shrinkage properties of concrete is the total water content of the concrete. As the water content increases, the amount of shrinkage increases proportionally. Large increases in the sand content and significant reductions in the size of the coarse aggregate increase shrinkage because the

Drying-shrinkage cracks such as these are often the result of improper jointing.

total water is increased. Calcium chloride admixtures and high-shrinkage aggregates also increase shrinkage.

Concrete expands and contracts with a rise and fall in temperature. Because it is not uncommon for temperatures to rise and fall as much as 40°F during a 24-hour period, such a temperature swing is sufficient to cause cracking if the concrete is restrained.

Other potential causes of cracking and even failure in concrete slabs are insufficiently compacted subgrades, soils susceptible to frost heave or swelling, and overloading (placing a greater weight on a concrete member than it is designed to bear). Cracking in concrete can be reduced significantly or eliminated by doing the following.

- Properly prepare the subgrade.
- Maximize the size and amount of coarse aggregates.
- Minimize the amount of mix water required for workability.
- Avoid calcium chloride admixtures.
- Use spray-applied evaporation retarders or plastic sheets to prevent rapid loss of surface moisture.
- Provide control joints at proper intervals (see table 3-2).

- Provide isolation joints to prevent restraint from adjoining elements of a structure.
- Prevent extreme changes in temperature.
- Use a 2- to 3-inch layer of sand on vapor barriers.
- If concrete must be placed directly on vapor barriers, use concrete with a low water content.
- Properly place, consolidate, finish, and cure concrete.

Cracks may also be caused by freezing and thawing of saturated concrete, alkali-aggregate reactivity, sulfate attack, or corrosion of reinforcing steel. However, cracks from these sources usually do not appear for many years. Proper mix design and selection of materials significantly reduces or eliminates the formation of cracks and deterioration related to these causes.

Discoloration

Surface *discoloration* of concrete slabs can be a cause for concern. Discoloration may appear as gross color changes in large areas of concrete, spotted or mottled light or dark blotches on the surface, or early light patches of efflorescence.

Studies determining the effects of various concreting procedures and concrete materials show that no single factor is responsible for all discoloration. Factors found to influence discoloration are calcium chloride admixtures, cement alkalis, hard-troweled surfaces, inadequate or inappropriate curing, variation of the water-cement ratio at the surface, and changes in the concrete mix. The use of calcium chloride in concrete may darken the surface.

Extreme discoloration may also result from attempts to hard-trowel the surface after it has become too stiff

Driveway discoloration due to substituting one cement for another in the middle of a job. The photo illustrates just one of several types of discoloration.

to trowel properly. Vigorously troweling a surface to progressively compact it can reach the point where the water-cement ratio is drastically decreased in localized areas. This dense, low-water-cement-ratio concrete in the hard-troweled area is almost always darker than the adjacent concrete.

Waterproof paper and plastic sheets used to moist-cure concrete containing calcium chloride are known to give a mottled appearance to flat surfaces, because it's difficult to place and keep the cover in complete contact with the surface. The places that are in contact will be lighter in color than those that are not.

Individual cements may differ in color. Thus, substituting one cement for another may change the color of concrete. The color of the sand has an effect on the color of the concrete. A high-strength concrete with a low-water-cement ratio is darker in color than low-strength concrete with a high-water-cement ratio.

Efflorescence is a deposit, usually white in color, that occasionally develops on concrete or masonry surfaces after construction is complete. Water in moist, hardened concrete dissolves soluble salts. Through evaporation or hydraulic pressure, this saltwater solution migrates to the surface where the water evaporates, leaving a salt deposit. Chemicals in concrete can also react with air to form efflorescence. For example, hydration of cement in concrete produces soluble calcium hydroxide, which can migrate to the surface and combine with carbon dioxide in the air to form a white calcium carbonate deposit.

If any of the conditions that cause efflorescence—water, evaporation, or salts—are not present, efflorescence will not occur. Efflorescence is harmless unless large accumulations of deposits develop within surface pores, which could cause spalling or flaking of the surface.

The discoloration of concrete can be avoided or minimized by doing the following.

• Avoid the use of calcium chloride admixtures.

• Use consistent concrete ingredients, uniformly proportioned from batch to batch.

• Place, finish, and cure concrete in a timely and proper manner.

To get rid of discoloration, the first (and usually the best) remedy is an immediate, thorough flushing with water. Permit the slab to dry, then repeat the flushing and drying until the discoloration disappears. If possible, use hot water. Acid washing using concentrations of weaker acids, such as 3 percent acetic acid (vinegar), lessens carbonation and mottling discoloration. Treating a dry slab with a 10-percent solution of caustic soda (sodium hydroxide) gives some success in blending light spots into a darker background. Muriatic acid (a dilute solution of hydrochloric acid) can be used to remove efflorescence. Harsh acids (concentrated solutions) should not be used, because they are dangerous to handle and can expose the aggregate.

Spalling

Spalling is a deeper surface defect than scaling. It often appears as circular or oval depressions on surfaces or as elongated cavities along joints. Spalls may be 1 inch or more in depth and 6 inches or more in diameter, although smaller spalls also occur. Spalls are caused by pressure or expansion within the concrete, bond failure in two-course construction, impact loads, fire, or weathering. Joint spalls are often caused by improperly constructed joints, and spalls may occur over corroded reinforcing steel.

Spalling can be avoided by doing the following when working with concrete.

• Properly design concrete constructions for the environment and anticipated service.

• Use proper concrete mixes and concreting practices.

• Use proper curing procedures to prevent uneven curing between the edges, joints, and the rest of the slab.

Low Spots

Low spots can affect slab drainage or serviceability if items placed on the slab need to be level. Low spots are often caused by poor lighting during placement and finishing, improperly set forms and screeds, damage to form and screed grade settings during construction, use of overly wet or variably wet concrete, and poor placement and finishing techniques.

Low spots can be avoided by observing the following instructions.

• Use a low-slump, low-water-content concrete mix.

• Work with adequate lighting.

• Check grades and levels frequently and fill in the low areas.

• Strike off with a straightedge that is straight and true.

CHAPTER 7

Maintenance and Repair of Concrete Surfaces

Concrete is one of the most durable and carefree surfaces used in building today. But, as with any building material, there is the occasional need for maintaining and repairing a concrete surface. The following information will help you repair and maintain your concrete surfaces and structures, both old and new, so they will provide you with years of value and enjoyment.

The major reasons for removing stains and cleaning concrete surfaces are either to improve the surface appearance or to prepare the surface for a surface treatment or concrete overlay. The methods used for each reason are described in the following paragraphs.

Most externally caused stains can be removed from concrete surfaces without difficulty, although sometimes it is necessary to repeat the treatment until the desired result is attained. Removal of old, long-neglected stains may require patience and, when the staining matter is not known, some experimentation. After removal, evidence of the stain may remain as a shadow because of the depth of its penetration into the pores of the concrete.

Removing Stains

Stains may be removed from concrete by either dry (mechanical) or wet (chemical or water) methods or a combination of them. Common dry methods are sandblasting, grinding, *scarifying*, and scouring. Care should be taken when using steel-wire brushes because they can leave metal particles on the surface that later may rust and stain the concrete.

Wet methods involve application of specific chemicals. According to the nature of the stain or the type of chemical used, the treatment acts in one of two ways: it dissolves the staining substance so it can be blotted up from the surface or driven more deeply into concrete; or, it bleaches or chemically changes the staining substance into a product that will not show.

The chemicals are either brushed on or applied as bandages or poultices. A bandage consists of layers of soft, absorbent, white cotton cloth soaked in chemicals and pasted over the stain. A poultice is a paste made with a solvent or reagent and some finely powdered, absorbent, inert material. Selection of the solvent or reagent depends on the type of stain. The inert material can be calcium carbonate (whiting), calcium hydroxide (lime), talc (talcum powder), fuller's earth, or diatomaceous earth.

Enough of the solvent or reagent is added to a small quantity of the inert material to make a smooth paste. The paste is spread in a 1/4- to 1/2-inch-thick layer onto the stained area with a trowel or spatula and allowed to dry. The solvent dissolves the staining substance and absorbs it into the poultice. There it migrates to the surface where the solvent evaporates and the stain is left as a loose, dried, powdery residue that can be scraped or brushed off. The chief advantage of a poultice is that it prevents the stain from spreading during treatment and tends to pull the stain out of the pores of the concrete.

Precautions

Cleaning procedures should be carefully planned. Materials such as glass, metals, or wood adjacent to the area to be cleaned should be adequately protected because they can be damaged by contact with some stain removers. No attempt should be made to remove stains until they are identified and the removal agent or method determined. If the stain cannot be identified, it is necessary to experiment with different bleaches or solvents on an inconspicuous area. The indiscriminate use of inappropriate products or improper application of the products can result in spreading the stains over larger areas or cause more unsightly, difficult-to-remove stains. Removing stains from old concrete sometimes leaves the area much

lighter in color than the surrounding concrete because surface dirt has been removed along with the stain.

Many chemicals can be applied to concrete without appreciable injury to the concrete surface, but strong acids or chemicals with a strong acid reaction should be avoided. Even weak acids may roughen the surface if left on for any length of time. The stain should be saturated with water before application of an acid solution so that the acid will not be absorbed too deeply into the concrete.

Hydrochloric acid (muriatic acid) diluted to a 10-percent solution is often used as a finishing treatment to remove all traces of the staining material. It may, however, leave a yellow stain on white concrete, so phosphoric acid is preferable for use on white concrete. All acid-treated surfaces should be thoroughly flushed with water.

Chemicals for removing stains from concrete can be obtained from sources such as commercial and scientific chemical suppliers, drugstores, hardware stores, supermarkets, or even some service stations.

CAUTION: Most chemicals used for cleaning concrete are hazardous and toxic and require adequate safety precautions. Skin contact and inhalation should be avoided. As a general precautionary rule, rubber or plastic gloves should be worn as well as chemical safety goggles. If used indoors, adequate ventilation must be provided. Storage and handling instructions printed on container labels should be followed. Unused portions that have been taken from the original containers should be discarded—never return them to their original containers. And be sure to never store chemicals in unidentified containers.

Some stains can be removed by more than one method. The most effective method in each case is best determined by trial and error. Before tackling large stains, a small quantity of the removing agent should be prepared and applied to an inconspicuous area to assess its effectiveness. It is advisable where possible to try a few different agents before making a choice. The effectiveness of the method on the sample area should not be judged until it has dried for at least 1 week.

Treatments for Specific Stains

Stains caused by different compounds require different treatments. The most common stains are described below along with some recommended treatments.

Bitumens. Bitumens (asphalts, tar, or pitches) have very good adhesion to concrete. Consequently, stains caused by them are very difficult to remove, especially if the bitumen has been allowed to penetrate deeply into the concrete surface. The degree of penetration depends on the type of bitumen. Before applying any treatment, scrape off any excess bitumen and scrub the surface with scouring powder and water.

Molten bitumen can be satisfactorily removed because it doesn't penetrate the concrete. Cool the bitumen with ice until it is brittle and chip off with a chisel. Scrub the surface with abrasive powder to remove the residue and rinse with clear water.

Bitumen emulsions are very small drops of bitumen dispersed in water. They do not penetrate the concrete very deeply. Scrub the stain area with scouring powder and water. Do not use solvents because they will increase the penetration of the stain and satisfactory removal will be impossible.

Cutback bitumen is a solution of bitumen in a solvent. It penetrates deeply into the concrete and is practically impossible to remove completely. The intensity of a cutback bitumen stain can be reduced by application of a poultice impregnated with toluene (toluol) or benzene (benzol), which is derived from coal tar and is not to be confused with benzine, which is derived from petroleum. After the poultice is removed, scrub the surface with scouring powder and water.

Sandblasting also can be used to remove cutback bitumen stains, but the sandblasting must be carried very deeply or the bitumen that has migrated into the concrete will give the surface a salt-and-pepper look.

Blood. Wet a bloodstain with clear water and cover it with a thin, even layer of sodium peroxide powder. **CAUTION:** Do not breathe any of the peroxide dust or allow it to come in contact with the skin as it is very caustic. Sprinkle the powder with water or apply a water-saturated bandage and allow it to stand for a few minutes. Wash with clear water and scrub vigorously. Next, brush a 5-percent solution of acetic acid (vinegar) on the surface to neutralize any alkaline traces left by the sodium peroxide. Rinse with clear water at the end of the treatment.

Caulking compounds. Scrape as much of the compound off the surface as possible and apply a poultice impregnated with denatured alcohol. Let stand until dry. After this treatment, most caulking compounds become brittle and can be brushed off easily with a stiff brush. Finally, wash the surface thoroughly with hot water and strong soap, trisodium phosphate (TSP), or a commercial scouring compound.

Chewing gum. The chemical composition of chewing gum varies from one manufacturer to another and the artificial colorings used are generally strongly staining. Consequently, chewing gum and chewing gum stains are very difficult to remove from concrete surfaces. First try the removal suggested for caulking compound stains.

Another way is to scrape off as much chewing gum as possible and then remove the rest with a solvent such as chloroform or carbon disulfide.

Coffee, tea, alcoholic beverages, and soft drinks. Coffee, tea, alcoholic beverages, and soft drink stains can be removed by applying a bandage saturated with 1 part glycerol (glycerin) diluted with 4 parts water. Two parts of isopropyl alcohol may be added to this mixture to hasten the removal action. The bleaches described for the removal of fire stains are also effective in removing stubborn coffee stains.

Copper, bronze, and aluminum. Runoff water from copper flashing and bronze fixtures usually leaves bluish green stains, although in some cases they are brown. To remove them, dry mix 1 part ammonium chloride (sal ammoniac) with 4 parts fine-powdered talc, whiting, or clay. Add ammonium hydroxide (household ammonia) and stir to make a smooth poultice. Place this over the stain, and leave until dry. Repeat the treatment as often as necessary and finally scrub well with clear water.

Aluminum stains appear as a white deposit that can be treated with diluted hydrochloric acid. Saturate the stained surface with water and scrub with a solution of 10 percent hydrochloric (muriatic) acid following the precautions given for chemical cleaning. Weaker solutions should be used on colored concrete to prevent a change in color. Rinse thoroughly with clear water to prevent etching of the surface and penetration of the dissolved aluminum salts into the concrete. Should this happen, the salts later may reappear on the surface as efflorescence.

Curing compounds. Generally, curing compounds will be worn off in a relatively short time by the abrasion from normal use or by natural weathering. However, if an accelerated treatment is required, or if the stained surface is not subjected to abrasion, the following procedures can be used.

Curing compounds have different chemical formulations. They may have a synthetic resin base, a wax base, a combination wax-resin base, a sodium silicate base, or a chlorinated-rubber base. The base of the curing compound should be identified before an attempt is made to remove it.

Curing compounds based on sodium silicate can be removed by vigorous brushing with clear water and a scouring powder. Wax, resin, or chlorinated-rubber curing compounds can be removed by applying a poultice impregnated with a solvent of the chlorinate hydrocarbon type such as trichloroethylene, or a solvent of the aromatic hydrocarbon type such as toluene. A mixture of 10 parts methyl acetone, 25 parts benzene, 18 parts denatured alcohol, and 8 parts ethylene dichloride also can be effectively used. Allow the poultice to stand for 30 to 50 minutes. Scrub the surface with clear water and a detergent at the end of the treatment.

Old stains can be best removed by mechanical abrasive methods such as light grinding or sandblasting.

Efflorescence. Compared with other stains, the removal of most types of efflorescence is relatively easy. Most efflorescence salts are water soluble and can be removed by pressure-water jet, light sandblasting, or dry brushing, followed by flushing with clean water. The removal of heavy accumulations or stubborn deposits may require washing the stains with a dilute solution of muriatic acid (1 part acid to 12 parts water) and scrubbing with a stiff-bristled brush. Dampen the surface with clean water before applying the solution. After scrubbing thoroughly, flush with clean water to remove all traces of the acid. Some commercial compounds are available to remove efflorescence.

Epoxies. Most solidified epoxies can be removed from small areas by burning off with a blowtorch. Adequate ventilation must be provided because black, acrid smoke will be given off. If a black stain remains, it can be treated as indicated for fire stains. Abrasive blasting is more appropriate for large areas.

Fire, smoke, and wood tar. Smoke is a difficult stain to remove. Scour the surface with powdered pumice or grit scrubbing powder to remove surface deposits and wash with clear water. Follow this with application of a poultice impregnated with a commercial sodium or potassium hypochlorite solution (for example, Clorox™ bleach).

Cover the poultice with a slab of concrete or a glass pane, making sure the poultice is pressed firmly against the stain. If the stain is on a vertical plane, devise a method to hold the poultice snugly in place; resaturate the poultice as often as necessary.

Fruit. Fruit stains are organic and can be removed by scrubbing with a solution of a synthetic powdered detergent and warm water.

Graffiti. A large number of commercially available products are suitable for removing spray paint and felt-tip pen markings from concrete surfaces. These products are generally effective also for removing crayon, chalk, and lipstick. The manufacturer's directions should always be followed. If satisfactory results are not obtained with the first remover applied, a second or third attempt with other products should be made. A single product may not remove both spray paint and felt-tip pen stains.

Grease stains. Grease does not penetrate into concrete, so scraping and scrubbing usually will remove it. Scrape off all excess grease from the surface and scrub with scouring powder, soap, TSP, or detergent. If staining persists, methods involving solvents are required. Use benzene, refined naphtha solvent, or a chlorinated hydrocarbon solvent such as trichloroethylene to make a stiff poultice. Apply to the stain and do not remove until the paste is thoroughly dry. Repeat the application as often as necessary. If required, scrub with strong soap, scouring powder, TSP, or detergents specially made for use on concrete. Rinse with clear water at the end of treatment.

Iron rust stains. Iron rust stains are easily recognizable by their rust color or their proximity to steel and iron in or on the concrete. Sometimes large areas are stained by a curing water that contains iron. The appearance of the concrete can be improved by mopping with a solution containing 1 pound of oxalic acid powder per gallon of water. After 2 or 3 hours, rinse with clear water, scrubbing at the same time with stiff brushes or brooms. Bad spots may require a second treatment.

Microorganisms. Organisms such as algae and lichens can be destroyed by the application of a sodium hypochlorite solution (5 percent), a household bleach solution, or by a 3- to 5-percent aqueous solution of copper sulfate.

Mildew. Prepare a solution of 1 ounce of laundry detergent, 3 ounces of TSP, 1 quart of laundry bleach, and 3 quarts of water. Apply to the area with a soft brush. Rinse with clear water after the treatment.

Oil stains. There are several methods for removing oil stains from concrete. The first method is to saturate the area with mineral spirits or paint thinner. Then cover the area with an absorbent material such as dry portland cement, talc, cat litter, fuller's earth, cornmeal, or cornstarch. Let stand overnight, then sweep away the cover. Repeat if necessary.

If an oil stain resists this first method, scrub with a TSP solution or bleach the surface with laundry bleach. If none of these methods completely removes the stain, allow the concrete to dry, then apply a paste prepared from benzol and a dry, powdered material. Leave the paste on the stain for an hour after the benzol has completely evaporated. Repeat if necessary.

Wet paint stains. Carefully soak up freshly spilled paint with an absorbent material such as paper towels or soft cloth. Avoid wiping the paint because that will spread the stain and drive the paint into the concrete. Immediately scrub the stained area with scouring powder and water. Scrubbing and washing should be continued until no further improvement is noted. Wait 3 days for the paint to harden, because paint removers or solvents used on wet paint, or films less than 3 days old, spread the stain and increase its penetration into the surface, making removal very difficult. After 3 days use one of the dried paint removal methods.

Dry paint stains. Scrape off as much as possible of the hardened paint, then apply a poultice impregnated with a commercial paint remover. Allow to stand for 20 to 30 minutes. Scrub the stain gently to loosen the paint film and wash off with water. Any remaining residue can be scrubbed off with scouring powder. Commercial paint removers are available in the form of a gel solvent. Test these on a small area on a trial basis.

Color that has penetrated the surface can be washed out with dilute hydrochloric (muriatic) or phosphoric acid. This treatment can be applied also to dried enamel, lacquer, or linseed-oil-based varnish. For shellac stains, the paint remover should be replaced by alcohol.

Old, dried paint films may require sandblasting or burning off with a blowtorch. Urethane varnishes are best removed by grinding or sandblasting.

Stains other than those discussed can be removed by experimenting with different bleaches or solvents on an inconspicuous area. The treated area should always be thoroughly scrubbed with clear water after the treatment so that no traces of the removing agent remain.

Cleaning Concrete Surfaces

The method selected for cleaning concrete preliminary to repairing depends on the extent of the work to be done. It may entail only a bucket and brush; but, it may also involve a hammer and chisel, water pressure, sandblasting, chemicals, or flames.

When faced with the decision to repair concrete, it's best to start by making a careful investigation, which may bring unexpected facts to light. Accurate diagnosis of the problem is essential for effective and successful cleaning. And take note of the hazards inherent in each cleaning method. For example, water and chemical cleaners can lead to problems caused by excessive moisture or unanticipated chemical reactions, sandblasting and flame cleaning will change the texture and appearance of the surface, and power tools can damage thin sections or remove more concrete than is desirable. The following tips will provide some guidance for selecting the least-damaging method.

High-Pressure Water Blasting

With the development of high-pressure water-jetting equipment, water can be used to effectively clean fully cured concrete and masonry surfaces. High-pressure water blasting relies on the force of the water rather than on abrasives. Most high-pressure washers available from tool rental companies attain pressures between 5,000 and 10,000 psi. The nozzle on a pressure washer can be adjusted from a fan spray to a pinpoint spray. It's best to use the fan spray because a pinpoint spray can cut a hole completely through concrete.

High-pressure washers can be used to remove many stains, loose paint, algae, efflorescence, and so on. It is an excellent method for cleaning concrete prior to painting or applying a sealer. After use, allow at least 10 days to 2 weeks for the concrete to dry before applying any finishes.

CAUTION: Because of the force of these washers, *never* direct the spray toward your body because the water can be driven into your skin and lead to serious complications. Also, wear goggles and protective clothing and keep others away from your work area.

Low-Pressure Water Spraying

Low-pressure water spraying employs no more force than is generated by a garden hose and spray nozzle hooked to the household water supply. In this method, only enough water is sprayed onto the surface to keep the deposits of dirt moist until they soften. Larger amounts of water are no more effective, and they might oversaturate a masonry wall and penetrate to the building interior. Cleaning should begin at the top of the structure so that surplus water runs down and presoftens the dirt below.

On some surfaces the softened dirt can be removed by hosing down the concrete, but usually it's necessary to assist removal with the gentle use of bristle brushes and nonferrous (for example, brass) or stainless steel wire brushes. The low-pressure water spray method is effective only when the dirt lies lightly on the surface or is bound to the wall with water-soluble matter.

Sandblasting

Sandblasting drives an abrasive grit at concrete or masonry surfaces to erode away dirt, paint, various coatings or contaminants, and any deteriorated or damaged concrete. Sandblasting changes the appearance of the concrete surface, leaving it with a rougher texture that may hasten the need for recleaning. Sandblasting also removes the edges and any sharp detail on moldings and ornaments. Even the flat surface of hard, polished aggregate will become scarred and dulled.

Although the sandblasting operation is not complicated, there are certain procedures and safety precautions that must be followed to ensure safe operations and a uniform surface. Therefore, a sandblasting contractor should be hired to do this work.

Chemical Cleaning

The materials used in chemical cleaning can be highly corrosive and toxic. You must use special equipment for their application and wear protective clothing. In addition, protection may be necessary for adjacent structures, lawns, trees, and shrubs. For these reasons, chemical cleaning should be left to the specialist. If you decide to do it yourself, carefully read all directions on the cleaner and follow them closely.

Chemical cleaners are water-based mixtures formulated for use on specific types of masonry and concrete. Most of them contain organic compounds called *surfactants* (surface-active agents) that work as detergents, allowing the water to penetrate the surface dirt more readily, hastening its removal. In addition, the mixtures contain a small amount of either acid or alkali to help separate the dirt from the surface.

Acid etching is often suggested as a satisfactory method for cleaning a concrete surface. Hydrochloric acid, also known as muriatic acid, is widely used because of its ready availability. Cleaning with proprietary compounds, detergents, or soap solutions generally requires the same procedure.

The procedure for cleaning concrete using a diluted acid solution is as follows.

CAUTION: Most chemicals used for cleaning concrete are hazardous and toxic and require adequate safety precautions. Skin contact and inhalation should be avoided. As a general precautionary rule, rubber or plastic gloves should be worn as well as chemical safety goggles. If used indoors, adequate ventilation must be provided. Storage and handling instructions printed on container labels should be followed. Unused portions that have been taken from the original containers should be discarded—never return them to their original containers. And be sure to never store chemicals in unidentified containers.

1. Mix a 10-percent solution of muriatic acid (1 part acid to 9 parts clean water) in a nonmetallic container by pouring the acid into the water. Stronger acid solutions may have to be used if the etching action is insufficient.

2. Mask or otherwise protect windows, doors, ornamental trim, and metal, glass, wooden, and stone surfaces from acid solutions.

3. Remove dust and dirt from the area to be cleaned and presoak or saturate with water.

4. Apply the acid solution to the damp surface with spray equipment, plastic sprinkling cans, or a long-handled stiff-fiber brush. Allow the solution to remain for 5 to 10 minutes. Nonmetallic tools may be used to remove stubborn particles.

5. Rinse thoroughly. Flush the surfaces with large amounts of clean water before they can dry. Acid solutions lose their strength quickly once they are in contact with cement paste or mortar; however, even weak, residual solutions can be harmful to concrete. Failure to completely rinse the acid solution off the surface may result in efflorescence or other damaging effects. Test with pH paper and continue until a pH of 7 or higher is obtained.

Before deciding on a particular method, clean a relatively small area to assess the efficiency of the method and the appearance and condition of the surface after the treatment. The reasons for cleaning must be considered carefully, because results with methods intended to improve only the appearance can differ substantially from results with methods to prepare the surface for a coating or concrete overlay.

Patching and Repairing Concrete

New or old structures sometimes require repairs that may range from patching minor surface imperfections such as tie holes and honeycomb areas to major structural renewal. Prior to patching any surface, loose or deteriorated concrete must be removed with a hammer and cold chisel and all loose particles and dust removed. Undercut existing concrete for a good bond.

Treatment of Bonding Surfaces

A high-quality bond of concrete or mortar to concrete surfaces is impor-

To repair broken concrete, begin by chiseling out all loose and weak concrete with a cold chisel and hammer. Undercut the sides of the repair area to "lock" the patch in place.

Use compressed air or hand bellows to blow all loose material out of the hole.

Apply a sand-cement grout or bonding compound to the old surface; then fill the crack with patching mortar or concrete and trowel it smooth.

tant in producing durable structures. Inferior bond is often responsible for failure of repairs to existing structures. Concrete surfaces to which concrete or mortar is to be bonded should be rough, clean, and dry before placing the new material.

Although good bond is obtained with both dry and damp base concrete, tests show that the best bond is developed with a dry base concrete. In hot weather the surface may be dampened by light water spray, but the new material should not be applied to a surface that is extremely wet or one that contains free water in hollows and rough areas.

Just before placing concrete patch material, a 1:1 sand and portland cement grout (mixed with water to the consistency of thick paint), or a commercially available *bonding agent*, should be applied to the base. A layer of grout about $1/8$ inch thick should be carefully and thoroughly brushed into the surface, and the new concrete should be placed before the bonding grout sets or dries out. Proprietary bonding agents should be applied according to manufacturing specifications.

Dry-Pack Mortar

Dry-pack mortar should be used for filling holes left by the removal of form ties, for narrow grooves cut for repair of cracks, and for filling honeycomb areas where lateral restraint can be obtained. Honeycomb areas should be cut out to a depth of at least 1 inch, slightly undercutting the edges or cutting at right angles to the surface. (Make sure there are no feather edges.) The surface should be cleaned with water and dried. A bonding grout composed of 1 part portland cement to 1 part sand with enough water for a paintlike consistency should then be thoroughly brushed on the surface.

Dry-pack mortar is usually mixed with 1 part portland cement to $2^{1/2}$ parts sand (by dry volume or weight). Where uniform color is important, white cement may be used; the amount sufficient to produce a color match can be determined by trial mixes, using varying proportions of gray and white cement. Use only enough water to produce a mortar that will pack into a ball when molded by hand. Too little water will not make a sound, solid pack; too much water will result in excessive shrinkage and a loose repair.

The area should be built up with mortar placed in layers of about $3/8$-inch compacted thickness. The surface of each layer should be roughened to improve bonding with the next layer. One layer can be placed immediately after another unless an unstable, plastic condition develops. If this occurs, stop for 30 to 40 minutes to permit hydration and hardening.

Commercial Patching Compounds

Perhaps the easiest way to patch holes and other surface defects in concrete is to use commercially available patching compounds. These compounds are premixed and are either wet (resin-based) or dry. Because of their convenience, you can use these to effect most any repair with ease. Be sure to carefully follow all manufacturer's instructions.

Replacement Concrete

Where the area to be filled is deeper than 4 inches, it is often practical to bond replacement concrete on the prepared surface. With reinforced concrete, the reinforcing steel should be free of rust scale and should be properly positioned. Where necessary, new reinforcement tied to the old should be provided.

Repairs to the side of formed concrete require special forms built with a chimney opening or side hopper through which concrete can be placed. Wet the area to be patched and allow the moisture to equalize for several hours to a damp, but not wet, condition. Apply a bonding grout immediately before placing concrete. The concrete should be of a sticky, plastic consistency and should be carefully rodded into place. Pressure should be applied to the fill concrete, and the form should be vibrated for wide cracks or holes. For convenience, use prepackaged concrete mixes for these small repairs and be sure to mix according to the manufacturer's instructions. As with all concrete and mortar work, repairs must be moist cured for 5 to 7 days.

Repairing Step Edges

When step edges become chipped or start to disintegrate, you can repair them without going through the exercise of total step replacement. Using a cold chisel and hammer, chip out all loose concrete. For larger areas you'll need to undercut the concrete slightly to create a good bonding surface. After removing the loose concrete, clean the surfaces and wet the area. Allow the moisture to equalize for a few hours before patching so that the surface is damp, but not wet.

To support the patch on the front surface of the step, cut a board to the height of the step. Place the board snugly against the face of the step and brace with concrete block. After the form is set, brush on a proprietary bonding agent or a cement-sand grout as described earlier in this chapter. Next, mix the patch mortar (1:2½ to 1:3 cement:sand), or use an epoxy-cement or latex-based patching mix and fill the void. Use an edger to compact and edge the repair then trowel the surface after removing the form.

Wedge a board in place against the repair area and apply a cement-sand grout. Fill in the broken area with fresh mortar or patching mix.

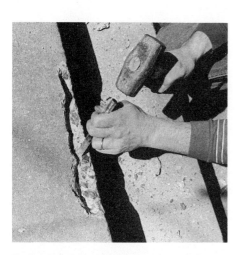

To fix the edge of a step, chip away all loose or deteriorated concrete and undercut it; then clean away all loose matter. Wet the area and allow it to stand before patching.

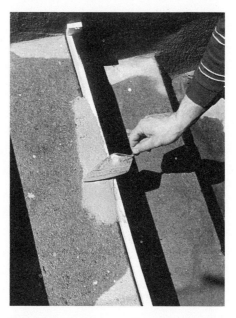

Trowel the mortar smooth, making sure the patched area is flush with the rest of the original step.

After placing the mortar patch, use an edger to compact and form it.

After the patch has set up, remove the form and use a trowel to smooth out the patch. Then cover the area to keep it moist while the patch cures.

Holes and Defects

After removing forms from new concrete, bulges, *fins*, and small projections may be present. These defects may be removed by chipping or tooling, then the surface should be rubbed or ground. Cavities or holes resulting from tie-rods should be filled unless they are to remain for decorative purposes. Usually honeycombed areas may also be repaired.

All of these defects can be minimized by exercising care in constructing the formwork and placing the concrete. In general, repairs for these defects should be made as soon as practical, preferably as soon as the forms are removed. Patches usually appear darker than the surrounding concrete; therefore, some white cement should be used in patching mortar or concrete where appearance is important.

Maintenance and Treatments for Concrete Floors

The durability of concrete floors depends primarily on observance of the fundamental rules in making, placing, finishing, and curing the concrete. Dusting of the floor surface may occur if these rules are violated.

Curing the freshly placed concrete is a very important step in securing dense surfaces that will not dust. The concrete should be kept wet for a minimum of 5 days in warm weather (70°F or higher) and 7 days in cooler weather (50° to 70°F). The concrete should not be permitted to dry out during the curing period.

Cleaning the New Floor

Floors become littered with plaster, mortar, and concrete droppings, and these materials are ground into the surface of the floor. When the floor is used, the layer of waste pulverizes under traffic, causing dust.

If the floor has not been protected during construction it should be given a thorough cleaning. After sweeping with an ordinary broom to remove loose dirt, use a vacuum cleaner to remove fine dust, then scrub the floor with soap and water and a wire brush. A final scrubbing should be given with soap and a fiber brush.

Hardener Treatments

If proper methods of construction have not been used and dusting occurs, it may be advisable to apply a chemical floor hardener to assist in hardening and binding the surface. These treatments are not cure-alls for poor materials or careless workmanship. They will improve many surfaces but will not make a perfect wearing surface of a poorly built floor.

It is essential that the floor be cleaned as described above before treatment. It should also be fairly dry to assist penetration.

Fluosilicate Treatment The fluosilicates of zinc and magnesium, when dissolved in water, have been used with good success. Either of the fluosilicates may be used separately but a mixture of 20 percent zinc and 80 percent magnesium appears to give the best results. In making up the solutions, 1 lb. of the fluosilicate should be dissolved in 1 gal. of water for the first application and 2 lb. to each gallon for subsequent applications. The solution may be mopped on or applied with a sprinkling can and then spread evenly with mops. Two or more applications should be given, allowing the surface to dry between applications. About 3 or 4 hours are generally required for absorption, reaction, and drying. Care should be taken to mop the floor with water shortly after the last application has dried to remove incrusted salts; otherwise white stains may be formed.

Sodium Silicate Treatment Commercial sodium silicate is a thick liquid and requires thinning with water before it will penetrate concrete. A good solution consists of 4 gallons of water to each gallon of silicate. Two or three coats should be used, allowing each coat to dry thoroughly before the next one is applied. Scrubbing each coat with a stiff-fiber brush and water will assist penetration of the succeeding application. Fluosilicates of zinc and magnesium have also been used with much success.

Coverage

The amounts of the above solutions required to treat floors will vary considerably with the porosity of the concrete. Generally, a gallon of any one of the solutions will be required for each application on 150 to 200 square feet of floor surface.

Slabjacking

Over time, slabs and foundations can lose subgrade support and undergo uneven settlement. Uneven settlement is often the result of flowing water undermining subgrades, inadequate soil consolidation, expansive and moisture-sensitive soils, or frost heave. A slab or foundation can be restored to its original elevation by *slabjacking*.

Slabjacking is merely "pressure grouting" or the process of pumping a flowable cement mixture under the element to be raised. The grout fills any voids, lifts the old concrete to the desired height, and stabilizes the raised concrete's location on hardening. Slabjacking is best performed by contractors who specialize in this work.

Tuckpointing

Tuckpointing is the act of replacing cutout or defective mortar joints with new mortar. There are two basic reasons why tuckpointing may be necessary: (1) there are leaks in the mortar joints and (2) the joints are deteriorating. Tuckpointing can produce a weathertight wall and help to preserve the structural integrity and the appearance of the masonry.

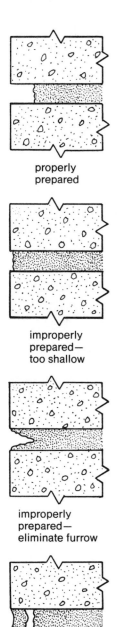

properly
prepared

improperly
prepared—
too shallow

improperly
prepared—
eliminate furrow

improperly
filled

If a wall is being tuckpointed to make it weathertight, it is recommended that all mortar joints in the wall get that treatment because minute cracks that could pass a visual inspection might still allow moisture to pass through. But, before beginning the job, make a thorough inspection of all flashings, lintels, sills, and caulked joints to ensure that water will not leak through into the masonry. If it is obvious that water is leaking through only one crack, it may be sufficient to tuckpoint only the mortar joints in the vicinity of that crack.

Preparation of Joints

Mortar joints should be cut out to a depth of at least ½ inch, and in all cases the depth of mortar removed should be at least as great as the thickness of the mortar joint. If the mortar is unsound, the joint should be cut deeper until only sound material remains. Shallow or furrow-shaped joints will result in poor tuckpointing (see illustration 7-1).

A tuckpointer's grinder with an abrasive blade is usually more efficient than hand chiseling for cutting out defective mortar. Use compressed air or a stream of water to remove all loose material.

The joints between clay masonry units should be dampened to prevent absorption of water from the freshly placed mortar. However, the joints should not be saturated just prior to tuckpointing because free water on the joint surfaces will act to impair bond. To avoid shrinkage, the joints between concrete masonry units should

Illustration 7-1. For successful tuckpointing, make sure joints are properly prepared and properly filled with mortar.

not be wetted before or during tuckpointing.

Preparation of Mortar

Tuckpointing mortar should have a compressive strength equal to or less than that of the original mortar or should contain approximately the same proportions of ingredients as the original mortar. (Recommended tuckpointing mortar mixes are shown in table 9-1.) Some masonry applications with structural concerns or severe frost or environmental conditions (such as horizontal surfaces exposed to weathering) may require the use of special mortars other than those shown in the table.

A recommended procedure for mixing tuckpointing mortar is as follows.

1. Mix all of the dry material, thoroughly blending the ingredients.

2. Mix in about half the water, or enough water to produce a damp mix that will retain its shape when formed into a ball by hand.

3. Mix the mortar for 3 to 7 minutes, preferably with a mechanical mixer.

4. Allow the mortar to stand for 1 hour for prehydration of the cementitious materials to reduce shrinkage.

5. Add the remaining water and mix for 3 to 5 minutes.

Tuckpointing mortar should have a drier consistency than conventional mortar for laying masonry units. Evaporation and absorption may require that water be added and the mortar remixed to regain proper workability. *Retemper* as needed; however, discard the mortar 2½ hours after the initial addition of water to the mix. Colored mortars may lighten with the addition of water; therefore, they should not be retempered.

Filling the Mortar Joints

The general method of applying mortar in joints that are to be tuckpointed is to use a *hawk* and a tuckpointing trowel. The hawk is used to hold a supply of mortar; it also catches mortar droppings if held against the wall just below the joint being filled.

The tuckpointing trowel should be narrower than the mortar joints that are being filled in order to obtain a proper degree of compaction. If the trowel does not fit into the joints it will be more difficult to obtain thoroughly compacted and completely filled joints.

Mortar should be spread into a joint in layers and firmly pressed to form a fully packed joint. Firm compaction is necessary to prevent voids as shown in illustration 7-1. The act of firmly compacting the mortar also helps ensure bond to masonry units and to the old mortar. Voids are undesirable because they may trap water, which can freeze and damage the new joint.

Tooling Mortar Joints

Tooling compacts the mortar to a dense surface with good durability. For weathertight construction, all mortar joints should be tooled to a concave or V-shape (see illustration 7-2). These shapes are recommended because they do not allow water to rest on the joint and they result in the mortar being pressed toward both the lower and upper masonry unit. This reduces weathering and helps ensure maximum bond between the mortar and the masonry units.

Jointing tools can be made from round or square bar stock. For horizontal joints, or vertical joints in the stacked bond pattern, the tool should be considerably longer than the masonry units to avoid a wavy joint.

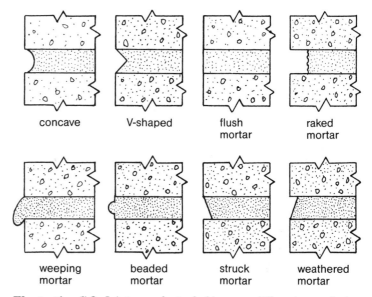

concave   V-shaped   flush mortar   raked mortar

weeping mortar   beaded mortar   struck mortar   weathered mortar

Illustration 7-2. Joints may be tooled in many different ways, but the concave and V-shaped are the most weather resistant.

Weeping, beaded, raked, flush, struck, and weathered joints (see illustration 7-2) can be used for decorative effects, but are not recommended for maximum weathertightness and durability. They can be satisfactorily used for interior work and for exterior work when the masonry is protected from the elements or located in climates that do not impose extremely low temperature or water saturation.

The cementitious materials in mortars require a period in the presence of moisture to develop proper strength. The mixing water in the mortar will usually provide this necessary moisture. However, freshly placed mortar should be protected from the sun and drying winds. With severe drying conditions, it may be necessary to cover the masonry structure with polyethylene sheeting or apply a fine water-fog spray for about 4 days to reduce evaporation of water from the mortar.

CHAPTER 8

Concrete Projects

Concrete is one of the most durable materials you can use for improvements around the home. Yet, many times it is only considered for major projects such as walks, steps, patios, and foundations. The truth is, concrete is also appropriate for many small, decorative objects that add a touch of friendliness to your lawn and garden. Many such items are available for purchase at garden and concrete specialty centers; but, in this chapter we will show you how to make some of them yourself. Once you see how simple it is, you won't hesitate to create your own variations on these designs.

Garden Bench

Lawn or garden benches lend an inviting, restful air to your outdoor living area. Concrete benches of many designs are sold by precast product producers. But, you can also make your own garden bench by following these simple instructions and the specifications shown in illustration 8-1.

1. Begin the bench by making a wooden form for the seat. Cut four pieces of 2 × 4 to length according to the drawings and fasten them together with double-headed nails.

2. Make the forms for the end supports, following the specifications shown in the drawings. (You can get by with one form by casting the legs one at a

Materials List

5 cu. ft. concrete	1:2$\frac{1}{2}$:2$\frac{1}{2}$:$\frac{1}{2}$ (cement:sand: gravel:water) or ten 60-pound bags of prepackaged mix
2 × 4s for forms	25′ (bench and two legs)
reinforcement	#3 rebar or 6″ × 6″, 6-gauge mesh
bolts (4)	$\frac{1}{2}$″ × 3″
bolts (4)	$\frac{5}{8}$″ × 3″
form release agent	silicone spray

time; but, two forms are more convenient.) Because the seat is anchored to the legs with bolts, drill two ⅝-inch-diameter holes through the upper end of each form, setting them 8 inches on center, as shown in the drawing. Insert greased ⅝ × 3-inch machine bolts to create holes in the cast concrete.

3. Cut and attach ⅝-inch-thick pieces of plywood to one side of each frame using galvanized drywall screws or wood screws of suitable length. For a decorative seat edge, tack cove molding to the inside of the seat form.

4. Apply form release agent to the frames. (Silicone spray is good for this.)

5. Mix the concrete for the bench, using gravel no larger than ¾ inch. See tables 2-3 and 2-4 for a suggested concrete mix. (Prepackaged concrete mixes are also satisfactory.) You will need 5 cubic feet of concrete to make the seat slab and the two end supports.

6. Place a 2-inch layer of concrete in the seat form. With a trowel work the concrete carefully along the inside edge so that it is tight against the form.

7. Level the concrete, then place the reinforcing material and ½ × 3-inch bolts on it as shown in the drawings. Add concrete up to the top of the forms and strike it off level with a 2 × 4, using a sawing motion. Check to make sure the bolts are vertical and the same distance apart as the bolts in the leg forms. Smooth the surface with a wooden float.

8. Allow the mixture to stiffen, and after the watery sheen is gone, use a steel trowel for the finish. Use an edging tool around the form to obtain a smooth, dense, well-rounded edge.

9. Place concrete in the leg forms level with their upper edges, then finish in the same manner as the seat.

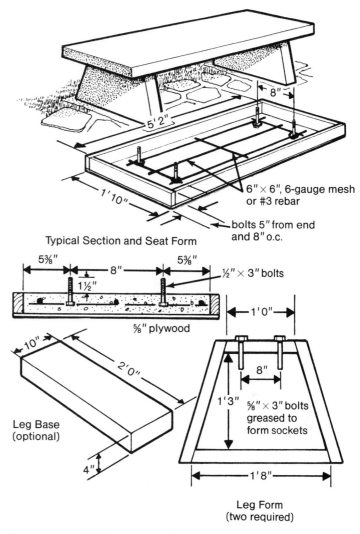

Illustration 8-1. Concrete garden bench.

10. If you wish, cast concrete bases for the legs to stand on. You can cast these bases directly in place or make forms for them as well. To cast-in-place, select the site for your bench and dig two 10-inch-wide × 24-inch-long × 4-inch-deep holes in the ground keeping the corners square and sides straight. Wet the soil thoroughly and allow the moisture in the soil to even out for several hours. Place the concrete in the holes. Make sure each slab finishes off level and that the two are level with each other.

11. Allow the concrete to cure in the forms for 24 hours then turn the slabs over and remove the frames (you'll need one or more helpers for this). Make a patching compound of cement and water mixed to a creamy consistency and patch any irregularities. Cover the pieces with plastic and allow to cure for 5 days, then set the legs in place and lower the bench top into position.

Decorative Block Mold

Even though there are numerous decorative block designs on the market, you may wish to create your own, build a mold, and cast your own blocks. The mold shown in illustration 8-2 features a herringbone pattern and produces a block 6 inches thick, 18 inches wide,

and 14 inches high. Once you understand the basic procedure, feel free to adapt the project to create blocks of different sizes, shapes, and patterns. Blocks of this type can be put to many uses. They can serve as legs for a garden bench, the ends of a planter box, or the walls of a charcoal grill, to name a few.

Here's how to make the mold shown in the drawing.

1. The mold wall is made from a 60-inch length of 2×3 stock. Set the piece on edge and rip it at a 15-degree angle, making the lower edge $1^3/8$ inches in width. Then cut the piece into two $11^1/2$-inch lengths and two $15^1/2$-inch lengths, mitering the ends so that the pieces form a rectangle that slopes in at the top. Fasten the pieces together using waterproof glue and 4d finish nails.

2. Cut a piece of $3/4$-inch plywood to 14×18 inches for the mold base. Center the mold wall over the base and fasten the two together using waterproof glue and a pair of $^{\#}6 \times 1^1/2$-inch flathead wood screws inserted through the base into each section of the mold wall.

3. Cut a piece of $3/4$-inch plywood to 16×20 inches for the carrying board. Turn the mold wall/base assembly upside down so you can center the carrying board over the base. Fasten the board to the base using waterproof glue and either $1^1/4$-inch underlayment nails or galvanized drywall screws.

4. Cut a piece of $3/4$-inch plywood to the dimensions needed to fit over the top of the mold wall (approximately $10^3/16 \times 14^3/16$ inches). While cutting, bevel the edges of the plywood to keep them in a continuous plane with the

---

Materials List

lumber for mold	$2'' \times 3'' \times 5'$
	$1'' \times 1'' \times 12'$
plywood for mold	$3/4'' \times 3' \times 4'$
fasteners	4d finish nails
	$^{\#}6 \times 1^1/2''$ flathead wood screws
	$1^1/4''$ galvanized drywall screws
	waterproof glue
	$4''$ loose pin hinges (4)
polyurethane varnish	1 pint
prepackaged sand-cement mix	60 lb. bag
form release agent	silicone spray

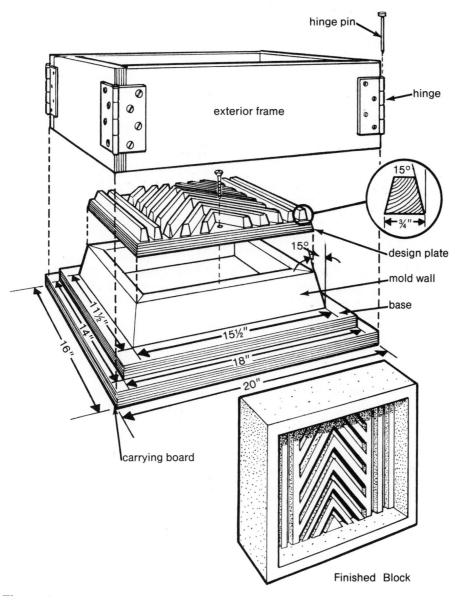

hinge pin

hinge

exterior frame

15°

¾"

design plate

15°

mold wall

base

11½"

14"

16"

15½"

18"

20"

carrying board

Finished Block

Illustration 8-2. Decorative block mold.

sides of the mold wall. This piece of plywood serves as the design plate.

5. Use a pencil and straightedge to lay out the pattern for the wooden strips to be attached to the top surface of the design plate. Allow $3/4$ inch for the width of each strip and $3/8$ inch for the space between adjacent strips. Lay out the location for the vertical strips on each end of the plate first, then square a line across the center of the plate to guide you in laying out the herringbone pattern between the end strips.

6. Cut the strips out of 1×1 stock, putting a 15-degree bevel on both sides and all exposed ends. Fasten the strips to the design plate using waterproof glue and 4d finish nails.

7. Place the design plate on top of the mold wall and drill pilot holes through the plate into the mold wall. Place the holes between the end strips near the four corners. Fasten the plate to the mold wall using #6 $\times 1\frac{1}{2}$-inch flathead wood screws. Countersink the screws so their heads will be flush with the surface of the plate.

8. Cut two pieces of $3/4$-inch plywood to $6^{3}/4 \times 14^{3}/4$ inches and two pieces to $6^{3}/4 \times 18^{3}/4$ inches for the exterior frame of the mold. Fasten the pieces together with their ends chasing each other using a 4-inch loose pin hinge at each corner. The resulting interior dimension of the frame should be 14×18 inches.

9. Putty all nail and screw holes, cracks, and imperfections on surfaces that will come into contact with concrete. Then sand those surfaces and coat them with three layers of polyurethane to seal the wood.

10. To use the mold, fit the exterior frame around the mold base and spray the inside with silicone to break bond. Add enough water to a bag of pre-packaged sand-cement mix to make a free-flowing mixture, then pour it into the mold. Use a wooden straightedge to strike off the mixture even with the top of the mold. After 24 hours, pull a pin from one of the hinges and remove the exterior frame. Wait another 24 hours before removing the mold. (You may have to tap it gently to loosen it.) Let the block cure under polyethylene film or wet burlap another 3 or 4 days before using.

Flower Boxes

Concrete flower boxes add an elegant look to any patio, and these flower boxes are sure to be a hit on your patio or even your entryway. Here are general guidelines for making flower boxes, as well as plans for a specific size and shape box (see illustration 8-3); but, feel free to adapt the design to suit your personal tastes.

1. In constructing a concrete flower box, you need both an inside and an outside form. Any width or length may be chosen to fit your plans. The flower box in the drawings utilizes standard lumber with beaded molding for patterns and decorative fluting. You can select screen molding or half-round molding for your boxes.

2. Construct the outside form first and tack the molding in place. Screw the sides together and the base to the sides using galvanized drywall screws or #8 $\times 1\frac{1}{4}$-inch flathead wood screws. A 1×1 wood strip offset at the top of the form gives the flower box a decorative lip.

3. Construct the inside form so you will have a planter with $1\frac{1}{2}$-inch-thick walls. This form does not break down after casting so nail the stock together using 8d galvanized common nails.

Attach the hanging bar across the top of the form. This bar must extend over the ends of the outside form in order to suspend the inside form.

4. Seal the forms with shellac or clear plastic spray, and use silicone spray as a form release agent.

5. Use white portland cement and white aggregate for a durable white finish. Gravel size should not exceed $3/4$ inch. The fresh concrete, mixed to the approximate proportions of $1:2\frac{1}{2}: 2\frac{1}{2}:\frac{1}{2}$ (cement, sand, gravel, water) must be well spaded in the forms. Tap the forms lightly as you work.

6. Cure the concrete by leaving the forms in place for 24 hours. After removing the forms, patch any irregular-

ities with a mixture of cement and water mixed to a creamy consistency. Cure for 5 to 7 days under a plastic sheet or wet burlap.

Materials List

form material	$3/4'' \times 4' \times 4'$ exterior plywood $1'' \times 1'' \times 8'$ lumber decorative molding
concrete	$1:2\frac{1}{2}:2\frac{1}{2}:\frac{1}{2}$ (cement:sand: aggregate:water)
fasteners	$^{\#}8 \times 1\frac{1}{4}''$ flathead wood screws 8d galvanized common nails
form release agent	silicone spray
form sealer	shellac or clear plastic spray

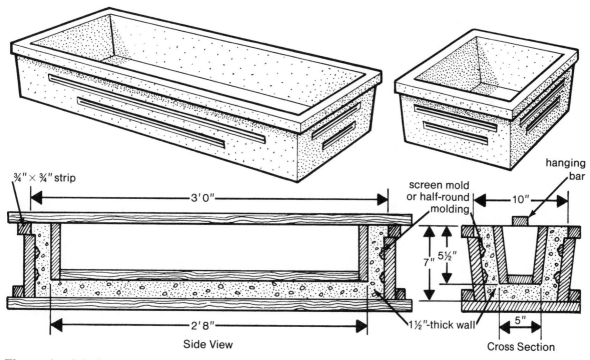

Illustration 8-3. Concrete flower boxes.

Post Projects

Concrete posts may be used for a variety of things around the home. Signs, clotheslines, mailboxes, and lamps are just a few of the many proj-ects you can make using sturdy, per-manent concrete posts. Each of these projects has some unique characteris-tics but the basic construction tech-niques are the same.

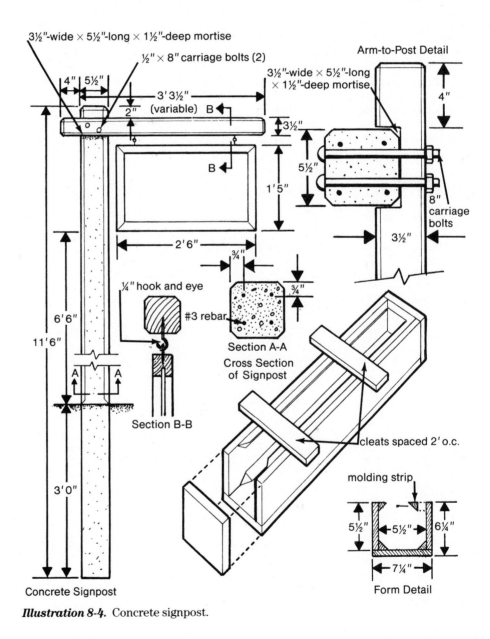

Illustration 8-4. Concrete signpost.

To create forms for $5^1/_2 \times 5^1/_2$-inch posts, fasten a pair of 1×6s to a 1×8 base, setting them $5^1/_2$ inches apart as shown in illustration 8-4. Every 2 feet, run cleats across the top of the form to brace the sides. For a decorative touch, tack cove or quarter-round molding into the corners of the form.

Signposts

Unique names of farms, ranches, and homes are an interesting part of Americana. Show off your special place with an attractive sign on a sturdy concrete post (see illustration 8-4). You may also want to consider combining a name and house number sign with an outdoor lamppost (see illustration 8-5).

The post described here is designed to be set 3 feet into the ground to balance the offset load of the sign and supporting arm. If you live in an area subject to heavy frost and you plant the post in ground with poor drainage, you should set it even deeper because of potential frost heave. Set it 6 inches below the prevailing frost depth in your area, but not less than 3 feet in any case.

1. Build the form using 1×6 material and optional decorative molding. Fasten the decorative strips only along the section of the form that will create the visible part of the post. Also attach the molding to the cap piece to create a beveled top.

2. Use a $1:2^1/_2:2^1/_2:^1/_2$ (cement:sand: aggregate:water) concrete mix. Use gravel of $^3/_4$-inch maximum size. White portland cement with white aggregate is often used to get a pure white post that needs no painting. Consider tinting the mix with an oxide of iron.

3. Seal the forms with clear plastic spray or shellac. Apply a silicone spray

form release agent, then cast about $^3/_4$ inch of concrete into the form and spade it. Lay two $^3/_8$-inch-diameter reinforcing rods on the concrete then place concrete to about $^3/_4$ inch below the top of the form. As you spade the concrete, be careful not to disturb the reinforcing rods. Gently rap the sides of the form to release any air pockets.

4. Place two more reinforcing rods on the concrete and cast the rest of the concrete. Form holes for the hanger arm by inserting oiled or shellacked $^3/_4$-inch-diameter *dowels* in the forms.

5. Strike off and float the concrete, then leave the form in place for at least 24 hours. After you strip the form, any imperfections in the post can be re-

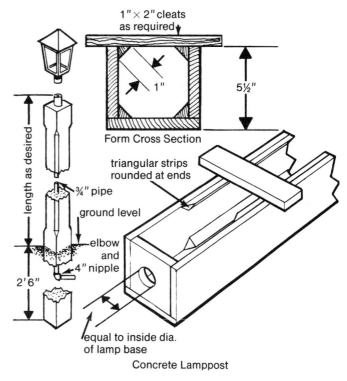

Illustration 8-5. Concrete lamppost.

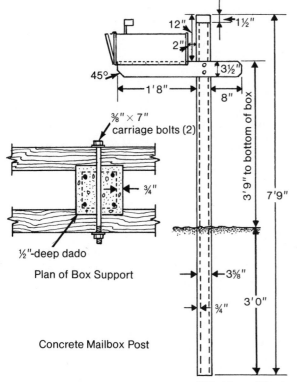

Plan of Box Support

Concrete Mailbox Post

Side View of Post

Illustration 8-6. Concrete mailbox post.

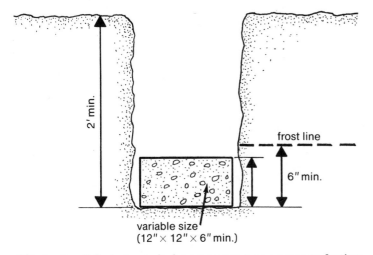

Illustration 8-7. A pier or deck post may rest on a concrete footing poured with or without a form below the frost line.

moved by working a cement-sand mortar into the imperfections with a wooden or sponge rubber float. Wrap the post with polyethylene or wet burlap and moist-cure for 5 to 7 days.

6. Cut a $3\frac{1}{2}$-inch-wide $\times 5\frac{1}{2}$-inch-long $\times 1\frac{1}{2}$-inch-deep mortise in the 4×4 crosspiece as shown; when the post is ready, assemble it with $\frac{1}{2} \times 8$-inch galvanized carriage bolts, nuts, and washers.

7. To make the sign frame, take an 8-foot length of 2×2 ($1\frac{1}{2}'' \times 1\frac{1}{2}''$) treated stock and rout a $\frac{1}{2}$-inch-wide $\times \frac{9}{16}$-inch-deep groove down the center of one face. Then cut this piece into two pieces 30 inches and two pieces 17 inches in length. Miter the ends of the pieces so that the grooved edges will be on the inside of the assembled frame.

8. Cut a piece of $\frac{1}{2}$-inch exterior plywood to 15×28 inches for the sign. Assemble the sign and caulk all cracks in the frame before priming and painting it. Mount $\frac{1}{4}$-inch-diameter hooks and eyes in the frame and crosspiece.

9. Erect the completed signpost by digging a 3-foot-deep hole and setting it. Fill the hole with dirt in 6-inch layers tamping each layer thoroughly before adding the next. Use a level to constantly check your work. Once the post is in place, hook the sign in place. Secure the sign by crimping the hooks, if desired.

Mailbox Posts

Many attractive concrete mailbox posts dot the countryside, showing the durable quality of concrete. The mailbox post is made with the same construction as the signpost except for the length, thickness of the post, and the crosspiece configuration (see illustration 8-6). Cut $\frac{1}{2}$-inch-deep dadoes in the 2 × 4s and cut angles on the ends of the boards as indicated in the drawings.

Mailbox flanges fit over the outside of the wooden supports. Your name, address, and box number painted on the box completes the project.

Deck Supports

Building a wooden deck onto your home not only adds value to your home but also provides you with a great place to enjoy the outdoors during warm weather. However, building a long-lasting deck that will provide years of service requires solid support and this is typically done with cast-in-place concrete footings or piers, or posts set in concrete. Here's how to proceed with each method.

Footings

Footings and piers are commonly used to support wooden decks and they can provide a sound foundation. Footings are usually placed at least 2 feet below ground level and no less than 6 inches below the frost line (this could be as much as 4 feet or more in some climates). The intent of footings is to supply a stable area for the weight of the deck to be distributed. The size of the footing should be between 8 to 12 inches square by at least 6 inches thick. (See illustration 8-7.)

Dig holes in the ground for the footings and if the earth is stable, cast the concrete directly into the ground using no forms. Be sure to wet the earth thoroughly and allow the moisture to equalize for a few hours prior to pouring the concrete.

Piers

Piers are built on top of footings to most any desired height. They can be below grade, grade level, above

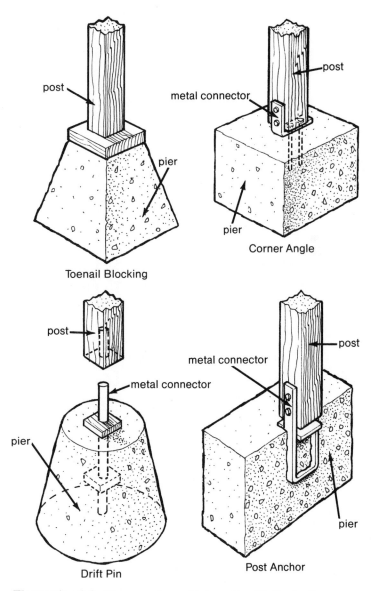

Illustration 8-8. Piers may be cast integrally with the footing or cast separately. Metal connectors provide the best means of anchoring a deck post to a pier.

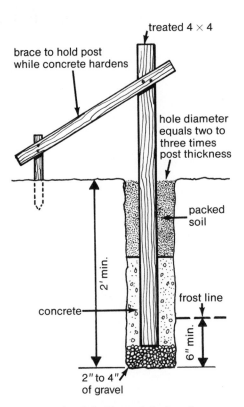

treated 4 × 4

brace to hold post
while concrete hardens

hole diameter
equals two to
three times
post thickness

packed
soil

2' min.

frost line

concrete

6" min.

2" to 4"
of gravel

Illustration 8-9. Treated deck or fence posts may be placed on gravel in a hole and concrete placed around them. Braces are used to keep the posts vertical while the concrete sets.

grade, or directly support the deck support beams. To make piers, build wooden forms, buy waxed cardboard forms, or stack concrete blocks and fill the cavities.

If you choose, you can end the piers 18 inches below grade then tamp the earth around 4 × 4 or 4 × 6 treated posts. If you elect to build piers to grade or above, embed galvanized metal connectors to secure deck posts with nails or bolts. Metal connectors not only secure the posts but also raise them off the pier, keeping them out of contact with moisture. (Optional ways of anchoring posts to piers are shown in illustration 8-8.)

Posts in Concrete

An economical method of stabilizing posts for a deck is to cast the post in concrete (see illustration 8-9). Begin by digging 12-inch-diameter holes to the same depth as discussed above for footings and piers. Place 2 to 4 inches of gravel in the holes, then set the treated posts on the gravel and pour concrete around them. If you are using prepackaged concrete mixes, keep in mind that the bags come in various sizes to make $1/3$, $1/2$, or $2/3$ cubic feet of concrete. Mix to a stiff consistency. Posts need to have braces tacked to them to keep them vertical while the concrete sets. This post-in-concrete method is recommended only for pressure-treated lumber used on small decks or wooden fences.

CHAPTER 9

Mortars for Masonry Walls

Centuries ago, combinations of sand and lime were used as mortar. These combinations took months and even years to harden as the lime slowly combined with carbon dioxide from the air to form calcium carbonate. Because it took so long for these mortars to harden and gain strength, it was necessary to use very thin joints. In many instances the joints were so thin that adjacent masonry units would bear on each other in direct contact. This type of construction required an excessive amount of labor to carefully fit and place each masonry unit. However, sand-lime mortars were adequate for the then massive construction and slow-paced construction procedures.

The development of mortars that hardened and gained strength rapidly made it possible to place masonry units more quickly. Also, thicker joints provided cushions for dimensional variations in the masonry units. The stronger mortars were first obtained by sweetening the lime with a small amount of portland cement. Later the amount of portland cement in the ratio was increased until the process involved sweetening the cement with a small amount of lime.

The development of strong mortars in the late 19th century was not the only great step forward for masonry construction. Another important step was the development of masonry cements. A masonry cement is a factory-prepared combination of materials that produces a mortar having the desired properties. Masonry cement includes portland cement; a plasticizing material, such as finely ground limestone, hydrated lime, or certain clays or shales; air-entraining agents; and sometimes water-repellent agents. White masonry cement and colored masonry cements containing premilled mineral oxide pigments are also available in many areas.

Since their development, masonry cement mortars have grown in popularity so that today most mortars are made with masonry cement. Masonry cements are designed to ease the mixing operation at the construction site and increase quality control by combining carefully selected materials into one package under factory control. Their use produces color and batch uniformity. Masonry cements that meet American Society for Testing and Materials (ASTM) standards ensure workable, sound, and durable mortar.

Mortar for concrete masonry is designed not only to join masonry units into an integral structure with predictable performance properties, but also to do the following:

• Effect tight seals between units against the entry of air and moisture

• Bond with steel joint reinforcement, metal ties, and anchor bolts, if any, so that they perform integrally with the masonry

• Provide an architectural quality to exposed masonry structures through color contrasts or shadow lines from various joint-tooling procedures

• Compensate for size variations in the units by providing a bed to accommodate tolerances of units

Masonry mortar is composed of one or more cementitious materials; clean, well-graded masonry sand; and sufficient water to produce a plastic, workable mixture. Modern specifications call for proportions by volume ranging from 1 part cementitious material to $2^1/_4$ to 3 parts of damp, loose mortar sand. The choice of cementitious material—masonry cement, a portland cement and lime combination, or a portland cement and masonry cement combination—is largely a matter of economics and convenience. Any of these combinations will produce mortar with acceptable properties as long as the specifications are met (see table 9-1).

Properties of Fresh Mortar

Good mortar is necessary for good workmanship and proper structural

---

Table 9-1. **Selection and Proportioning of Mortar**

Building Segment	Mortar Type	Parts by Volume				
		Portland Cement	Masonry Cement Type		Hydrated Lime	Sand
			S	N		
Exterior locations at or below ground level such as foundation walls, retaining walls, pavements, walks, and patios	S	$1/_2$ — 1	— 1 —	1 — —	— — $1/_2$	4 3 4
Interior walls and above-ground exterior walls and tuckpointing	N	— 1	— —	1 —	— 1	3 6

NOTE: Sand is measured in a damp, loose condition and 1 cu. ft. of masonry sand by damp, loose volume is considered equal to 80 lb. of dry sand.

performance of concrete or clay masonry. Because mortar must bond masonry units into strong, durable, weathertight walls, it must have the properties described below.

Workability

Probably the most important quality of a masonry mortar is *workability* because of its influence on other important mortar properties in both the plastic and hardened states. Workability is difficult to define because it is a combination of a number of interrelated properties.

An experienced mason judges the workability of mortar by the way it adheres to or slides from a trowel. Mortar of good workability should spread easily on the masonry unit, cling to vertical surfaces, extrude readily from joints without dropping or smearing, and permit easy positioning of the unit without subsequent shifting due to its weight or the weight of successive courses. Mortar consistency should change with the weather: a good workable mix should be softer in summer than in winter to compensate for water loss.

Water Retention

Mortar having good water retention resists rapid loss of mixing water (prevents loss of plasticity) to the air on a dry day or to an absorptive masonry unit. Rapid loss of water causes the mortar to stiffen quickly, making it practically impossible to obtain good bond and weathertight joints.

Water retention is an important property and is related to workability. A mortar that has good water retention remains soft and plastic long enough for the masonry units to be carefully aligned, leveled, plumbed, and adjusted to proper line without dan-

ger of breaking the intimate contact or bond between the mortar and the units. When low-absorption units such as split block are in contact with a mortar having too much water retention, they may float. Consequently, the water retention of a mortar should be within tolerable limits.

Entrained air, extremely fine aggregate or cementitious materials, or water add workability (plasticity) to the mortar and increase its water retention.

Consistent Rate of Hardening

The rate of hardening of mortar due to hydration (chemical reaction) is the speed at which it develops resistance to an applied load. Too rapid hardening may interfere with your use of the mortar. Very slow hardening may impede the progress of the work because the mortar will flow from the completed masonry. During winter construction, slow hardening may also subject mortar to early damage from frost action. A well-defined, consistent rate of hardening assists you in laying the masonry units and in tooling the joints to the same degree of hardness. Uniform color of masonry joints reflects proper hardening and consistent tooling times.

Hardening is sometimes confused with a stiffening caused by rapid loss of water, as when low-water-retention mortar is used with highly absorptive units. Also, during very hot, dry weather mortar may tend to stiffen more rapidly than usual. In this case, you may find it advisable to lay shorter mortar beds and fewer units in advance of tooling.

Properties of Hardened Mortar

There are also a number of properties to take into account in assessing the quality of a mortar once it has been placed and has hardened. We will begin our survey of these properties, beginning with the most important one of all, which is the *bond* that the mortar creates between masonry units.

Bond

The term bond refers to a specific property that can be subdivided into the extent of bond (the degree of contact of the mortar with the masonry units) and tensile bond strength (the force required to separate the units). A *chemical* and a *mechanical* bond exist in each category.

Good extent of bond (complete and intimate contact) is important to watertightness and tensile bond strength. Poor bond at the mortar-to-unit interface may lead to moisture penetration through the unbonded areas. Good extent of bond is obtained with a workable and water-retentive mortar, good workmanship, full joints, and masonry units having a medium initial rate of absorption (suction).

Tensile bond strength is perhaps the most important property of hardened mortar. Mortar must develop sufficient bond to withstand the tensile forces brought about by structural, earth, and wind loads, expansion of clay brick; shrinkage of concrete masonry units or mortar, and temperature changes.

Bond is affected by the following variables:

• Mortar ingredients, such as the type and amount of cementitious materials, the air content, and the amount of water retained

• Characteristics of the masonry units, such as surface texture, suction, and moisture content

• Workmanship, such as the pressure applied to the mortar bed during placing

• Curing conditions, such as temperature, relative humidity, and wind

All other factors being equal, mortar bond strength is related to mortar composition, especially the cement content. The bond strength of the mortar increases as the content of cement increases.

Bond is low on smooth, molded surfaces, such as glass or die skin surfaces of clay brick or tile. On the other hand, good bond is achieved on concrete block or on wire-cut or textured surfaces of clay brick. The suction rates of concrete masonry units are low enough that they never require wetting prior to laying of mortar. Some clay brick units have such high suction rates that poor bond will result unless the brick are wetted; surfaces of the wetted brick should be dry before applying mortar.

There is a distinct relationship between mortar flow (water content) and tensile bond strength. For all mortars, bond strength increases as water content increases. The optimum bond strength is obtained by using a mortar with the highest water content compatible with workability, even though mortar compressive strength decreases.

Workmanship is paramount in ensuring bond strength. The time lapse between the spreading of mortar and the placing of the masonry units should be kept to a minimum because the water content of the mortar will be reduced through suction of the masonry unit on which it is first placed. If too much time elapses before a unit is placed, the bond between the mortar and the unit will be reduced. You should not realign, tap, or in any way move units after initial placement, leveling, and alignment. Movement breaks the bond between unit and mortar, after which the mortar will not readhere well to the masonry units.

Portland cement requires a period in the presence of moisture to develop its full strength. In order to obtain optimum curing conditions, the mortar mixture should contain the maximum amount of water compatible with acceptable workability. Freshly laid masonry should be protected from the sun and drying winds. With severe drying conditions, it may be necessary to cover the masonry with a plastic sheet.

Durability

The durability of masonry mortar is its ability to endure the exposure conditions. Although aggressive environments and use of unsound materials may contribute to the deterioration of mortar joints, major destruction is from water entering the masonry and freezing.

In general, damage to mortar joints and to mortar bond by frost action has not been a problem in masonry wall construction above grade. In order for frost damage to occur, the hardened mortar must first be water saturated or nearly so. After being placed, mortar becomes less than saturated due to the absorption of some of the mixing water by the units. The saturated condition does not readily return except under special conditions, such as the following.

- The masonry is constantly in contact with saturated soils.
- Downspouts leak.
- There are heavy rains.
- Horizontal ledges in the mortar joints are formed.

Under these conditions the masonry units and mortar may become satu-rated and undergo freeze-thaw deterioration.

High-compressive-strength mortars usually have good durability. Because air-entrained mortar will withstand hundreds of freeze-thaw cycles, its use provides good insurance against localized freeze-thaw damage. Mortar joints that have deteriorated due to freezing and thawing present a maintenance problem that generally must be remedied by tuckpointing.

Strength

The principal factors affecting the compressive strength of masonry structures are the compressive strength of the masonry unit, the proportions of ingredients within the mortar, the design of the structure, the workmanship, and the degree of curing. Of these factors, compressive strength of mortar is largely dependent on the type and quantity of cementitious material used in preparing the mortar. Strength increases with an increase in cement content and decreases with an increase in air entrainment, lime content, or water content.

Shrinkage

A popular misconception is that mortar shrinkage can be extensive and cause leaky structures. Actually, the maximum shrinkage across a mortar joint is minuscule and therefore not troublesome. This is more frequent with the weaker mortars. They have more ability to bend or move without cracking and so are better able to accommodate shrinkage.

Appearance

Uniformity of color and shade of the mortar joints greatly affects the overall appearance of a masonry structure. Atmospheric conditions, admixtures, and moisture content of the masonry units are some of the factors affecting the color and shade of mortar joints. Others are uniformity of proportions in the mortar mix, water content, and time of tooling the mortar joints.

Careful measurement of mortar materials and thorough mixing are important to maintain uniformity from batch to batch and from day to day. Control of this uniformity becomes more difficult with the number of ingredients to be combined at the mixer. Pigments, if used, will provide more uniform color if premixed with a stock of cement sufficient for the needs of the whole project. In some areas, colored masonry cements are available.

To ensure a uniform mortar shade in the finished structure, it is necessary to tool all mortar joints at a similar degree of setting. If the joint is tooled when the mortar is relatively hard, a darker shade results than if the joints are tooled when the mortar is relatively soft. Some masons consider mortar joints ready for tooling once the water sheen is gone and the mortar has begun to stiffen, but is still thumb-print soft.

White cement mortar should never be tooled with metal tools because the metal will darken the joint. A glass or plastic joint tool should be used. (Tooling techniques are described in Chapter 11.)

Types of Mortar

Mortar type is identified through various combinations of portland cement with masonry cement, masonry cement singly, and combinations of portland cement and lime.

The amount of water to be used on the job is the maximum that will produce a workable consistency during construction. This is unlike conventional concrete practice where the water content must be carefully controlled. It should be noted that no one mortar type will produce a mortar that rates highest in all desirable mortar properties. Adjustments in the mix to improve one property often are made at the expense of others (see box 9-1).

The selection of mortars for masonry depends on the type of structure being built (see table 9-1). The choice of using masonry cement or a portland cement and lime combination is largely a matter of economics and convenience. Masonry cements provide all cementitious materials required for a masonry mortar in one bag. The quality and appearance of mortars made from masonry cement are consistent because the masonry cement materials are mixed and ground together before being packaged. Consequently, masonry cement mortars are less subject to variations from batch to batch than mortars produced from combining ingredients on the job.

Prepackaged Dry Mortar Mixes

For most home projects, you can buy prepackaged mortar mix. This dry-batched product overcomes the prob-

---

Mortar for Stone

The formula for mortar for stonework is a little different than for brick or concrete masonry. The mixture is richer and consists of 1 part portland cement to 3 or 4 parts sand. To this mixture you may want to add ½ part fireclay for bulk because so much more mortar is required in stonework. The main difference, though, is that you do not add any lime to the mix (and this includes not using prepackaged mortar mixtures that contain lime); lime may stain the stone. Also, because of the weight and density of stone, it is good practice to use less water in the mix in order to make it stiffer.

lem of adjusting the mix for moisture content of the sand and the usual problem of consistent portions of sand and cementitious materials. In prepackaged mixes, the cementitious materials and oven-dried sand are accurately weighed, mixed, and bagged. All you do is add the proper amount of water.

Colored Mortars

Pleasing architectural effects with color contrast or harmony between masonry units and joints are obtained through the use of white or colored mortars. White mortar is made with either white masonry cement plus white sand or with white portland cement and lime plus white sand. For colored mortars, the use of white masonry cement or white portland cement instead of the normal gray cements not only produces cleaner, brighter colors but is essential for making pastel colors such as buff, cream, ivory, pink, and rose.

Integrally colored mortar may be obtained through use of color pigments, colored masonry cements, or colored sand. Brilliant or intense colors are not generally attainable in masonry mortars. The color of the mortar joints will depend not only on the color pigment, but also on the cementitious materials, aggregate, and water-cement ratio.

Pigments must be thoroughly dispersed throughout the mix. To determine if mixing is adequate, some of the mix is flattened under a trowel. If streaks of color are present, additional mixing is required. For best results, the pigment should be premixed with the cement.

As a rule, color pigments should be of mineral oxide composition. Iron, manganese, chromium, and cobalt oxides have been used successfully (see table 9-2). Zinc and lead oxides

Table 9-2. Guide to Oxides for Tinting Mortar

Desired Color	Oxide
Red, yellow, brown, black	iron oxide
Green	chromium oxide
Blue	cobalt oxide
Black or gray	carbon or iron oxide

should be avoided because they may react with the cement. Carbon black may be used as a coloring agent to obtain dark gray or almost black mortar, but should be limited to 3 percent by weight of the portland cement. Lampblack should not be used at all.

When mixing pigment with mortar, add only the minimum amount needed to produce the desired shade. An excess of pigment, more than 10 percent of the portland cement by weight, may be detrimental to the strength and durability of the mortar. The quantity of water used in mixing colored mortar must also be accurately controlled, because the more water, the lighter the color. Avoid adding water to retemper the mix during use. Any colored mortar not used while plastic and workable should simply be discarded.

Fading of colored mortar joints may be caused by efflorescence, the formation of a white film on the surface. The white deposits are caused by soluble salts that have emerged from below the surface, or by carbonate compounds created when calcium hydroxide liberated during the setting of the cement combines with carbon dioxide in the air. Good color pigments do not effloresce or contribute to the problem. If efflorescence occurs, remove it with a stiff-bristled brush.

Efflorescence is a white film that can discolor both plain and colored masonry walls.

Ready-Mix Mortar

Ready-mix mortar in bags, pails, or tubs is available from some concrete suppliers. It contains all the ingredients, including water, premixed to a ready-to-use consistency. A special admixture keeps the mortar from hardening for up to 72 hours. The mortar begins to harden after it is placed between masonry units.

Components of Mortar

Using the right materials for your mortar is the first step in assuring you will get the best results. Mortar consists of cementitious materials such as portland cement, masonry cement, and hydrated lime; masonry sand; and water. When buying cement, lime, or sand, suppliers will more than likely help you make the best selection for your needs.

Water intended for use in mixing mortar should be clean and free of soluble salts (soluble salts can contribute to efflorescence later). Generally, for good results, water used for mixing mortar should be fit to drink.

Sand used in mortar must be specified as mortar sand. Make sure the sand you buy has been washed.

Measuring the Materials

Mortar ingredients should be measured in such a way as to ensure uniformity of mix proportions, yields, workability, and mortar color from batch to batch. Aggregate proportions are generally expressed in terms of loose volume, but experience has shown that the amount of sand can vary due to moisture bulking. In order to get the most consistent results when using mortar sand, protect the sand from rain and ground moisture by storing it on plastic or concrete and covering it with plastic.

Aside from the sand, other mortar ingredients are often sold in bags labeled only by weight. Because mortar is proportioned by volume, it is necessary to know the volume equivalents (see table 9-3).

The usual practice of measuring sand by the shovelful can result in excessive oversanding or undersanding of the mix. For more positive control, the following method is suggested. Construct one or two wooden boxes 12 inches square and 6 inches deep and use them to measure the sand required in a batch. Add the cement or lime by the bag. Then add water, measuring by the pail. When the desired consistency of mix is determined, mark down the number of units of sand you used and fill a plastic 5-gallon pail with the equivalent. Mark the level on the bucket and use the bucket for the rest of the batches.

Mixing Mortar

To obtain good workability and other desirable properties of plastic

Table 9-3.	Unit Weights of Mortar Ingredients	
Ingredient		**Unit Weight, lbs. per cu. ft.**
Portland cement		94
Masonry cement		70
Hydrated lime (dry)		40
Hydrated lime (putty)		80

masonry mortar, the ingredients must be thoroughly mixed. Mortar can be mixed either by machine or by hand depending on the amount of mortar required for the job. Because mortar is generally only workable for about 2 hours, mix only enough to last for that amount of time.

Except possibly on very small jobs, mortar should be machine-mixed. A typical mortar mixer has a capacity of 4 to 7 cubic feet. Conventional mortar mixers are of rotating spiral or paddle-blade design with tilting drum. After all batched materials are together, they should be mixed from 3 to 5 minutes. Less mixing time may result in non-uniformity, poor workability, low water retention, and less than optimum air content. Longer mixing times may adversely affect the air contents of those mortars containing air-entraining cements, particularly during cool or cold weather. Longer mixing times may also reduce the strength of the mortar.

Batching procedures will vary with individual preferences. Experience has shown that good results can be obtained when about three-fourths of the required water, one-half of the sand, and all of the cementitious materials are briefly mixed together. The bal-

ance of the sand and the remaining water is then added. The amount of water added should be the maximum that can be tolerated and still attain satisfactory workability. Mixing is carried out most effectively when the mixer is loaded to its design capacity. Overloading can impair mixing efficiency and mortar uniformity. The mixer drum should be completely empty before loading the next batch.

When hand mixing of mortar becomes necessary, such as on small jobs, all the dry materials should first be mixed together by hoe, working from one end of a mortar box (or wheelbarrow) and then from the other. Next, two-thirds to three-fourths of the required water is mixed in with the hoe and the mixing continued as above until the batch is uniformly wet. Additional water is carefully added with continued mixing until the desired workability is attained. The batch should be allowed to stand for approximately 5 minutes and then thoroughly remixed with the hoe.

Retempering

Fresh mortar should be prepared at the rate it is used so that its workability will remain about the same

When you require at least 4 cubic feet of mortar in a 2 hour period, mix the mortar with a power mortar mixer for best results.

throughout the day. Mortar that has been mixed but not used immediately tends to dry out and stiffen. However, loss of water by absorption and evaporation on a dry day can be reduced by wetting the mortarboard and covering the mortar in the mortar box, wheelbarrow, or tub.

If necessary to restore workability, mortar may be retempered by adding water; thorough remixing is then necessary. Although small additions of water may slightly reduce the compressive strength of the mortar, the end effect is acceptable. Masonry built using plastic mortar has a better bond strength than masonry built using dry, stiff mortar.

Mortar that has stiffened because of hydration should be discarded. Because it is difficult to determine by sight or feel whether mortar stiffening is due to evaporation or hydration, the most practical method of determining the suitability of mortar is on the basis of time elapsed after mixing. As stated

To restore workability, mortar may be retempered.

earlier, mortar should be used within about 2 hours after mixing. If colored mortar is used, no retempering should be permitted. Additional water may cause a significant lightening of the mortar.

Grout

Grout and mortar are used differently, have different characteristics, and are handled differently. Thus, the two are not interchangeable.

Grout is an essential element of reinforced masonry. In reinforced, load-bearing masonry wall construction, grout is usually placed only in those wall spaces containing steel reinforcement. The grout bonds the masonry units and steel so that they act together to resist loads. In some reinforced, load-bearing masonry walls, all cores (with and without reinforcement) are grouted to further increase the wall resistance to loads. Grout is sometimes used to fill a portion or all of the cores in a nonreinforced, load-bearing masonry wall construction to give it added strength. (Grout may be ordered from a ready-mix concrete supplier.)

The fineness or coarseness of a grout is selected on the basis of the size of the grout space as well as the height of the space to be filled (see table 9-4.) For fine grout (grout without coarse aggregate), generally the smallest space to be grouted should be at least $3/4$ inch wide, such as that which occurs in the collar joint of two-wythe wall construction (the joint between inner and outer *wythes*).

Admixtures

Practice has shown that a grouting-aid admixture may be desirable when the masonry units are highly absorbent. The desired effect of the grouting aid is to reduce early water loss to the

Grout is used in both reinforced and nonreinforced masonry. In reinforced masonry, grout secures the vertical and horizontal steel reinforcement.

Table 9–4. **Grout Proportions**				
	Parts by Volume			
Type	**Portland Cement, Portland Blast-Furnace Slag Cement, or Portland-Pozzolana Cement**	**Hydrated Lime or Lime Putty**	**Aggregate Measured in a Damp, Loose Condition**	
			Fine	**Coarse**
Fine grout	1	0 to $1/10$	$2^{1}/_{4}$ to 3 times the sum of the volumes of the cementitious materials	—
Coarse grout	1	0 to $1/10$	$2^{1}/_{4}$ to 3 times the sum of the volumes of the cementitious materials	1 to 2 times the sum of the volumes of the cementitious materials

masonry units, to promote bonding of the grout to all interior surfaces of the units, and to produce a slight expansion sufficient to help ensure complete filling of the cavities. The use of calcium chloride is strongly discouraged in grout because of possible excessive corrosion of reinforcement, metal ties, or anchors.

Slump

All grout should be of fluid consistency but only fluid enough to pour without segregation. It should flow around the reinforcing bars and into all joints of the masonry, leaving no voids. There should be no bridging or honeycombing of the grout.

The consistency of the grout, as measured using a slump test, should be based on the rate of absorption of the masonry units and temperature and humidity conditions. The desired slump is 8 inches for units with low absorption and up to 10 inches for units with high absorption.

Batching and Mixing

Whenever possible, grout should be batched, mixed, and delivered in accordance with the requirements for ready-mix concrete. Because of its high slump, ready-mix grout should be continuously agitated after mixing until placement.

Batching and mixing of grout on the job site is usually not recommended unless unusual conditions exist that would require special consideration or if only small quantities are required (see table 9-4). When a batch mixer is used on the job site, all materials should be thoroughly mixed for a minimum of 5 minutes. Grout not placed within $1^1/_2$ hours after water is first added to the batch should be discarded.

Curing

The high water content of the grout and the partial absorption of this water by the masonry units will generally provide adequate moisture within the masonry for curing both the mortar and grout. In dry areas where the masonry is subjected to high winds, some added protection (such as plastic sheeting) may be necessary.

Grouts placed during cold weather are particularly vulnerable to freezing during the early period after grouting because of their high water content.

CHAPTER 10

Masonry Unit Construction

Mortar is used as a sort of glue to hold together masonry units, and these units usually consist of one of three types of material—concrete, clay brick, or stone. Stone is undoubtedly the oldest of all building materials, followed by brick, which has been around for a few thousand years. By contrast, concrete masonry is a relatively new building material (the first concrete block was manufactured in 1882). Yet, the popularity of this material has developed to the point that concrete masonry units comprise more than two-thirds of the masonry in walls being built today.

Masonry units manufactured today are highly engineered construction materials, built to meet very precise standards, specifications, and codes. As can be seen from the photographs and drawings in this chapter, different types of masonry units have been developed for different purposes: in addition to conventional building blocks, there are special blocks for use in control joints, channel bond blocks made to receive rebar, decorative blocks for use in screen walls, and blocks with various unusual surface textures.

Not only are there masonry units available for every purpose, there is also a wide variety of mortar, grout, and finishes available for use with concrete masonry units. In the pages that follow, we will explore both the rich diversity of materials available and the basic techniques for working with them.

Concrete Masonry Units

Concrete masonry units (block and concrete brick) are available in sizes, shapes, colors, textures, and profiles for practically every conceivable need and convenience in masonry construction. In addition, concrete masonry units may be used to create attractive patterns and designs to produce an almost unlimited range in architectural treatments of wall surfaces. The list of current applications is lengthy, but

131

some of the more prominent uses are for:

- Exterior load-bearing walls (below and above grade)
- Interior load-bearing or non-load-bearing walls
- Fire walls, party walls, curtain walls
- Partitions, panel walls, solar screens
- Backing for brick, stone, stucco, and other exterior facings
- Veneer or nonstructural facing for wood, concrete, or masonry
- Piers, pilasters, columns
- Bond beams, lintels, sills
- Retaining walls, slope protection, ornamental garden walls
- Chimneys and fireplaces (indoor and outdoor)
- Paving and turf block

Concrete masonry units are manufactured in the United States to conform to certain requirements put forth by the American Society for Testing and Materials (ASTM). Table 10-1 describes two grades and two types of concrete unit.

The primary difference between the two grades is the density of the material. This property affects construction, insulation, acoustics, appearance, porosity, and the kind of finish that the material will take. The density of the material is also closely related to the amount of water it will absorb. Absorption, in turn, affects the quality of mortar needed. If a masonry unit absorbs water fast, the mortar will need to retain more water. This is necessary to give you time to place and adjust the block before the mortar stiffens and to achieve a strong mortar bond.

What you need to determine are the conditions to which the materials you purchase will be subjected. Sales staff at reputable building supply yards can then see that you get the right type of masonry unit and the appropriate mortar for the project you have in mind.

Manufacture

Concrete masonry units are mainly made of portland cement, graded aggregates, and water. Depending on specific requirements, the concrete mixtures may also contain other suitable ingredients such as an air-entraining agent or coloring pigment. Mass production techniques used by the industry make the cost of quality concrete masonry units relatively low.

Briefly, the manufacturing process involves the machine-molding of very dry, no-slump concrete into the desired shapes, which are then subjected to accelerated curing. This is generally followed by a storage or drying phase so the moisture content of the units may be reduced to the specified

Table 10-1. **Grades of Concrete Masonry Units for Various Uses (United States)**		
Grade	**Block**	**Concrete Brick***
N	for general use such as in exterior walls below and above grade that may or may not be exposed to moisture penetration or the weather and for interior walls and backup	for use as architectural veneer and facing units in exterior walls and for use where high strength and resistance to moisture penetration and severe frost action are desired
S	limited to use above grade in exterior walls with weather-protective coatings and in walls not exposed to the weather	for general use where moderate strength and resistance to frost action and moisture penetration are required

*Also applicable to solid concrete veneer and facing units larger than brick size, such as split block.

moisture limits prior to shipment. The concrete mixtures are carefully proportioned and their consistency controlled so that texture, color, dimensional tolerances, and other desired physical properties are obtained. High-strength units are made from concrete with higher cement contents and more water, but still no slump.

For storage or shipment to the building site the units are generally placed in small stacks, or *cubes*, consisting of layers of 15 to 18 units per layer. The cubes are stacked on wooden pallets or banded with the bottom layer of block having horizontal cores. Most delivery trucks are equipped with a device for unloading the cubes at the building site.

Types

Concrete masonry units are available in a wide variety of weights, sizes, shapes, and exposed surface treatments for virtually any architectural and/or structural function. Manufacturers and their local distributors can supply literature describing available types of units. It is important to determine what is available before proceeding with any construction plans calling for anything but the most common units.

Normal-Weight and Lightweight Block   The terms *dense,* or *normal-weight,* and *lightweight* are derived from the density of the aggregates used in the manufacturing process. The heaviest normal-weight blocks weigh between $1^1/_2$ and 5 times as much as lightweight blocks. In addition to density, the weight of an individual concrete block depends on the volume of concrete in the unit. Where weight of the construction material is of concern, the lightweight blocks should be considered, otherwise, normal weight is recommended.

Hollow and Solid Units   Concrete block are classified as *hollow* or *solid* units. A hollow unit is defined as one in which the net concrete cross-sectional area parallel to the bearing faces is less than 75 percent of the gross cross-sectional area. Units having net concrete cross-sectional areas of 75 percent or more are called solid units. Most units range from 50 to 70 percent depending on the size and shape of the cores. Hollow units are usually preferred over solid units because of their reduced weight and ease of handling.

Solid masonry units are used primarily for special needs, such as for the top or bearing course of load-bearing walls. Concrete brick and some split-block units are made 100 percent solid, although some concrete brick designs include a shallow depression, called a *frog,* in one bearing face. The purpose of the frog is to reduce weight, provide for better bond, and prevent the unit from floating when laid in the wall.

Modular Sizes   The dimensions of concrete masonry units are for the most part based on some module, usually 4 or 8 inches. From common usage the $^3/_8$-inch-thick mortar joint has become standard. Accordingly, the exterior dimensions of modular units are reduced by the thickness of one mortar joint ($^3/_8$ inch). Thus, when laid in mortar, modular units produce wall lengths, heights, and thicknesses that are multiples of the given module. This permits you to plan building dimensions and wall openings that will minimize the expense of cutting units on the job.

It is common practice in specifying concrete block to give the wall or block width first, the course (or block) height second, and the block length third, followed by the name of the unit. The dimensions used are nominal rather than actual (see illustration 10-1).

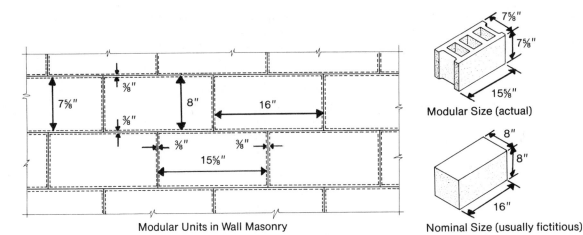

Modular Units in Wall Masonry

Modular Size (actual)

Nominal Size (usually fictitious)

Illustration 10-1. To allow for ³/₈-inch-thick mortar joints, dimensions for concrete block are given in nominal inches, which are ³/₈ inch larger than their modular dimensions.

The nominal block size that dominates the industry is $8 \times 8 \times 16$ inches. Other common dimensions are widths of 2, 4, 6, 10, and 12 inches. For convenience, half-length units (8 inches) and half-height units (4-inches) are available as companion items for completing various patterns. In some areas the 4-inch module is popular with nominal unit lengths of 8, 12, 16, 20, and 24 inches.

Basic Block Design Many masonry unit design variations are available. Some manufacturers make either two- or three-core units exclusively, while others make some sizes and shapes in both core designs, with the balance of their production in only one or the other core design.

There are variations in core and end designs as applied to $8 \times 8 \times 16$-inch concrete masonry units (see illustration 10-2). Some producers make regular *stretcher* units with flanged ends in the 8-, 10-, and 12-inch widths. Others have adopted, for regular production, single and double plain-ended designs, thereby meeting needs for stretcher, corner, or pier units from

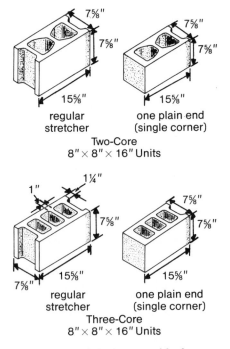

regular
stretcher

one plain end
(single corner)

Two-Core
8" × 8" × 16" Units

regular
stretcher

one plain end
(single corner)

Three-Core
8" × 8" × 16" Units

Illustration 10-2. Concrete block are generally made with 2 or 3 cores and with plain or grooved ends. Webs are flared for easier removal from molds, but the flares also provide for easier gripping when constructing walls.

single stocks of a given width. In general, all 4- and most 6-inch-wide hollow units are made with plain ends but may contain either two or three cores.

The cores of hollow units are tapered to permit ready stripping in the molding process. Some core designs also include a degree of flaring of the face or web to give a broader base for mortar bedding and for better gripping when handling. The face shells are sometimes thickened for greater tensile strength at these areas.

Size and Shape Variations When designing and building with concrete block, you may choose from a large number of sizes and shapes. A complete listing is not possible here as some plants produce several hundred different items. Some sizes and shapes are limited to certain areas or are made only on special order. Because it's expensive to stock all shapes, an individual supplier will probably have only a limited number of them.

In addition to the basic shapes, there are several unusual shapes commonly used in conventional construction. (Some of these are shown in illustration 10-3.) Half-length units are generally available in most shapes. Alternately, an easily split slotted or kerfed two-core unit may be used, or a masonry saw may be used to cut special shapes or shorter lengths from whole units.

Corner blocks have one flush end for use in pilaster, pier, or exposed corner construction. *Bullnose* blocks have one or more small-radius, rounded corners and are used instead of square-edged corner units to minimize chipping. *Jamb* or sash blocks are used to

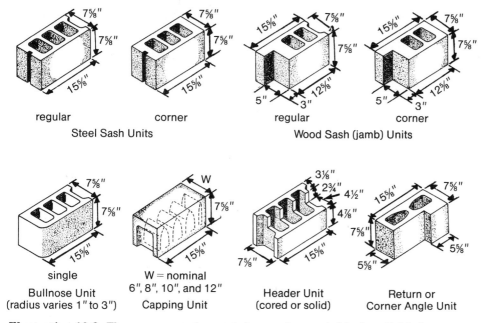

Illustration 10-3. There are many sizes and shapes of concrete block available for conventional wall construction. Dimensions shown here are actual and used for modular construction with ³/₈-inch joints. Half-length units are generally available in the shapes shown. Widths range from 4 to 12 inches (nominal), in 2-inch increments for some shapes.

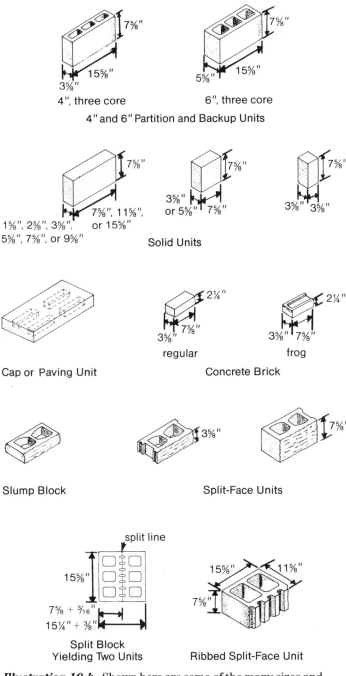

7⅝"

15⅝"

3⅝"

4″, three core

7⅝"

5⅝"   15⅝"

6″, three core

4″ and 6″ Partition and Backup Units

7⅝"

7⅝", 11⅝", or 15⅝"

1⅝", 2⅝", 3⅝", 5⅝", 7⅝", or 9⅝"

7⅝"

3⅝" or 5⅝"   7⅝"

7⅝"

3⅝"   3⅝"

Solid Units

Cap or Paving Unit

2¼"

3⅝"   7⅝"

regular

2¼"

3⅝"   7⅝"

frog

Concrete Brick

Slump Block

3⅝"

7⅝"

Split-Face Units

split line

15⅝"

7⅝ + ³⁄₁₆"

15¼" + ⅜"

Split Block Yielding Two Units

15⅝"   11⅝"

7⅝"

Ribbed Split-Face Unit

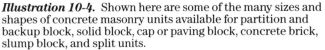

Illustration 10-4. Shown here are some of the many sizes and shapes of concrete masonry units available for partition and backup block, solid block, cap or paving block, concrete brick, slump block, and split units.

facilitate the installation of windows or other openings. *Capping* blocks have solid tops for use as a bearing surface in the finish course of a wall. *Header* blocks have a recess to receive the header unit in a composite masonry wall. *Return* or corner-angle blocks are used in 6-, 10-, and 12-inch-thick walls at corners to maintain horizontal coursing with the appearance of full-length and half-length units.

A number of block types are available for partition construction and facing unit backup (see illustration 10-4). Solid units and cap or paving units are manufactured in a variety of sizes for use as capping units for parapet and garden walls and for use in walkways, patios, fireplaces, barbecues, and as wall veneer. They may be used both structurally and nonstructurally. When they are used in reinforced walls, the reinforcing steel is generally placed in grout spaces between wythes.

Concrete brick are sized to be laid with ³⁄₈-inch mortar joints, resulting in modules of 4-inch widths and 8-inch lengths. The thickness of mortar joints is increased slightly so that three courses (three brick and three *bed joints*) lay up 8 inches high.

Slump units are produced by using a concrete mixture finer and wetter than usual. The concrete brick or block unit is squeezed to give a bulging effect. The rounded or bulging faces resemble handmade *adobe*, producing a pleasing appearance.

Split brick or split block are solid or hollow units that are fractured (split) lengthwise or crosswise by machine to produce a rough stonelike texture. The fractured faces, which are exposed when the units are laid, are irregular but sharp. The aggregates in the units are exposed in various planes of fracture. A wide variety of interesting colors, textures, and shapes are produced through the use of diverse

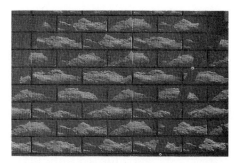

Slump-block units have rounded or bulging faces that resemble adobe.

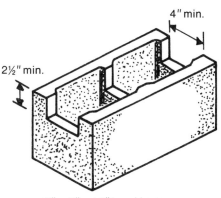

$8'' \times 8'' \times 16''$ bond beam

Split brick or block are solid or hollow units that are fractured to expose a rough, stonelike surface.

cements, aggregates, color pigments, and unit sizes.

The minimal length of split units is 16 inches, but half-length units, return corners, and other special units made in multiples of 4 inches are obtainable. The solid units are nominally 4 inches wide and available in various modular heights ranging from $1^{5}/_{8}$ to $7^{5}/_{8}$ inches. Split solid units are used as a veneering or facing material. Ribbed hollow units, which can be split to produce unusual effects, are useful for through-the-wall applications indoors or out. Specialized shapes for constructing windowsills, *copings*, bond beams, and lintels are also available (see illustration 10-5).

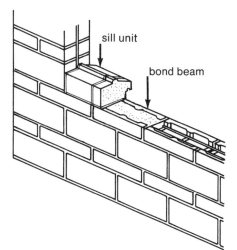

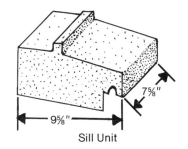

Sill Unit

Illustration 10-5. Specialized blocks are used for sills, copings, bond beams, and lintels.

Channel bond-beam blocks are made especially for steel-reinforced masonry walls.

In areas of recurring earthquakes or hurricanes, where major damage to construction has a high probability of occurrence, reinforced bond beams as well as vertical and horizontal wall reinforcement are mandatory. Reinforced lintels are necessary to bridge over openings for windows and doors. Various sizes and shapes of lintels are made to meet different requirements of load capacity, spans, wall widths, and window or door types. Specialized units are also available for constructing pilasters and columns (see illustration 10-6).

Screen block or grille units have gained wide popularity as decorative and functional masonry. The units available (see illustration 10-7) have a wide range of size, from 4 to 16 inches square, to meet nearly every need. Though often used mainly for their aesthetic value, they also provide an excellent balance between privacy and vision from within or without. They diffuse strong sunlight, provide a wind break, and yet permit free flow of air. The decorative value of the units is enhanced by the effects that variations in light and shade produce on their patterns. They are used principally to create decorative building facades, ornamental room dividers and partitions, garden fences, and patio screens.

Although a wide range of designs with screen block can be obtained, basically the designs can be divided into the following categories:

• Designs made with units that are a complete pattern in themselves

• Designs made with units that are combined in pairs or in fours

• Overall patterns created from units of several different designs

• Interesting designs created by leaving spaces between conventional solid or hollow block units

Architectural concrete masonry units, sometimes known as customized masonry or sculptured units, offer opportunities for almost unlimited architectural freedom (see illustration 10-8). Patterns and profiles in the block can be achieved with vertical scoring, fluted or ribbed faces, molded angles or curves, projected or recessed faces, or combinations of these surfaces.

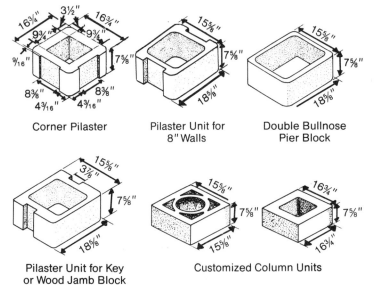

Corner Pilaster

Pilaster Unit for 8″ Walls

Double Bullnose Pier Block

Pilaster Unit for Key or Wood Jamb Block

Customized Column Units

Illustration 10-6. A variety of concrete masonry shapes is manufactured for pilasters, piers, and columns.

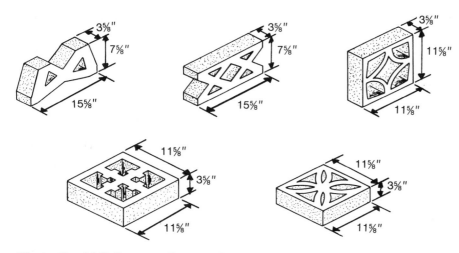

Illustration 10-7. Screen walls are made with units that form either all or part of a pattern.

Screen walls are aesthetically appealing and also provide an excellent balance between privacy and vision from both sides.

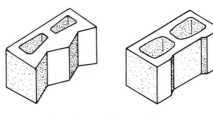

Scored, Ribbed, and Fluted Faces

Recessed Faces

Angular and Curved Faces

Illustration 10-8. Architectural concrete masonry units are produced in a wide variety of patterns. As with all concrete units, the available types vary with the supplier and the locality.

Prefaced Blocks

Prefaced concrete masonry units offer opportunities for a wide range of colors, patterns, and textures. Prefaced units are sometimes supplied with scored or patterned surfaces in a variety of colors. The color may be even or dappled, and the surface may be smooth or textured. A unit may be prefaced on one or two sides and on one end, as required. They also may be produced in a variety of thicknesses, heights, and shapes (such as the

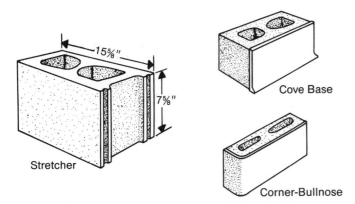

Illustration 10-9. Although used mostly in commercial applications, prefaced units are available in different shapes from local suppliers.

stretcher, bullnose, and cove base units shown in illustration 10-9).

Although prefaced block are commonly used in commercial construction, you may find applications for them around the home. But, be sure to check with your local building materials supplier on cost and availability before designing a project around block of an unusual design.

Surface Texture

The surface texture of concrete masonry units may be greatly varied to satisfy aesthetic requirements or to suit a desired physical requirement. Various degrees of smoothness can be achieved with any aggregate by changes in the aggregate grading, mix proportions, mix wetness, and the amount of compaction in molding. Textures are classified somewhat loosely and with considerable overlapping as open, tight, fine, medium, and coarse.

Fine, medium, and coarse describe the relative smoothness or graininess of the texture. A fine texture is not only smooth but made up of small, very closely spaced granular particles.

A coarse texture is noticeably large-grained and rough, resulting from the presence of a large proportion of large-sized aggregate particles in the surface. Usually, but not necessarily, it will contain substantial-sized voids between aggregate particles. A medium texture is one intermediate between fine and coarse.

An open texture is characterized by numerous closely spaced and relatively large voids between the aggregate particles. Conversely, a tight texture is one in which the spaces between aggregate particles are well filled with cement paste; it has few pores or voids of the size readily penetrated by water and sound.

If the concrete masonry surface is to serve as a base for stucco or plaster, a coarse texture is desirable for good bond. Coarse and medium textures provide sound absorption even when painted. The paint, however, must be applied in a manner that does not close all of the surface pores; therefore, spray painting is best. A fine texture is preferred for ease of painting.

Because of the methods used to manufacture block, absolute uniformity of texture is not possible. Since the general texture differences are a result of individual manufacturing processes and mixes between block producers, selecting a single source for your block is the best way to guarantee some degree of uniformity.

Ground-facing is an example of one special block texturing process. Ground-face masonry units are produced by grinding off a 1/16- to 1/8-inch layer of concrete from one or both face shells of regular block. The process results in a smooth, open-textured surface that shows aggregate particles of varying color to good advantage. Variations in aggregate size, type, and color and the use of integral pigments cre-

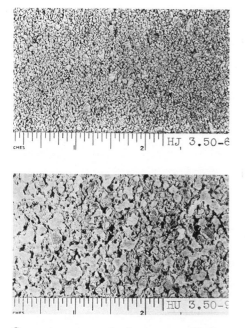

Ground-face units expose the natural color of the aggregate.

Concrete masonry textures vary with the types of aggregate used during the manufacturing process. The top unit was made with fine aggregate and the bottom with coarse aggregate.

ate many interesting texture and color options. Ground-face units in natural or tinted colors are often used in partition walls that are to be exposed without further finish, except perhaps the application of wax or a colorless sealer.

Color

The natural color of concrete masonry varies from light to dark gray to tints of buff, red, or brown, depending on the color of the aggregate, cement, and other mix ingredients used as well as the method of curing. Because of these factors, colors may vary.

Units also are subject to temporary and permanent changes in color. Colored surfaces are more vivid and darker when wet than when dry. Units made with dark-colored aggregate will slowly become darker with age when subject to weathering because the surface film of cement paste erodes away exposing the aggregate. Storage of block at the plant, supplier, or at the job site affects color, but not substantially. As with texture, the best way to guarantee uniform color is to secure all your block from a single source. Of course, if the block is to be painted or covered with a surface finish, color is no real concern when you make your purchase.

Brick

Bricks are made out of clay that has been mixed with other earthen materials that alter its properties either in the raw or finished state. Water is added to the *dry mix* and then this mix is fed into a machine that compacts and removes air from the clay as it extrudes the shape. The extruded shape is cut into bricks and these bricks are then stacked, dried, and fired to the appropriate hardness.

Types and Sizes

There are two general categories of brick—common brick and face brick. Common brick are used for general building and vary from brick to brick in color. By contrast, face brick are uniform in color and are used where this is required. Within these two groups there is an almost unlimited variety of size (see illustration 10-10), texture, and color. There are also both solid and hollow-core varieties. Cores reduce the weight of a brick and en-

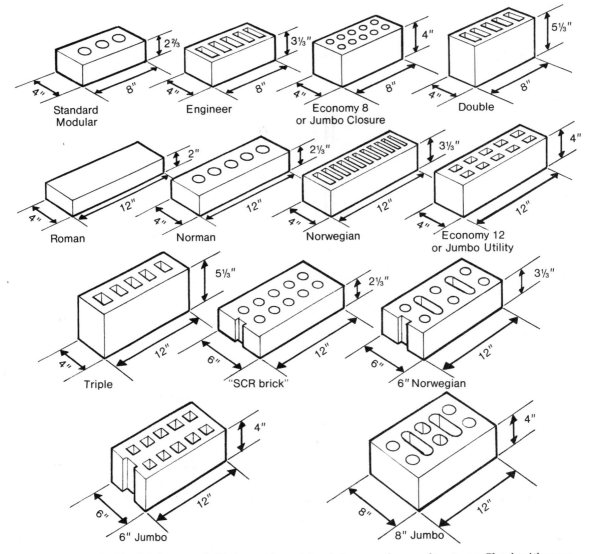

Illustration 10-10. Clay brick are available in a wide variety of shapes, colors, and textures. Check with your local supplier for specific types and sizes.

hance the mortar bond, so cored brick are often used in wall construction.

A third type is firebrick, which is composed of special clays fired to temperatures higher than those used in the manufacture of common or face brick. Firebrick are bonded with a special fireclay mortar and are used in fireplaces because of their enhanced ability to withstand thermal heat shock.

The sizes of brick are usually given in nominal dimensions. The standard size is $4 \times 2^2/_3 \times 8$, a nominal measurement that includes the mortar joint thickness. When using brick for a particular project, you will need to know the method of construction you are going to use and the dimensions of the project in order to select the right number of brick. Firebrick are usually larger than common brick and typically have real dimensions of $4^1/_2 \times 2^1/_2 \times 9$.

Stone

Stone is the oldest of all building materials and its use in projects adds a natural warmth and feel to any home. Stone is available in two forms—natural or cut. Natural stone, also referred to as *rubble*, varies extensively in size and requires some skill in fitting and cutting; but, rubble is cheap or even free for the taking. Cut stone, also known as *ashlar*, is usually cut into squared-off units for easy manipulation. The most common variety of cut stone is veneer stone, which is generally cut to a 4-inch thickness. Veneer stone may readily be used as a facing on either frame or block construction (see illustration 10-11).

Masonry Walls

The design of a concrete masonry wall depends on its required appearance, economy, and strength. But the layout of the wall involves other important considerations, such as the internal arrangement of components, modular planning, weather resistance, and provisions for shrinkage cracking control. All deserve careful planning if the wall is to successfully serve its intended purpose.

Concrete masonry walls may be classified as solid, hollow, cavity, composite, veneered, reinforced, or grouted. These classifications sometimes overlap, but the basic terminology and bonding directions remain the same (see illustration 10-12).

Solid Masonry Wall

Solid masonry walls (see illustration 10-13) are built of solid masonry units with all joints completely filled

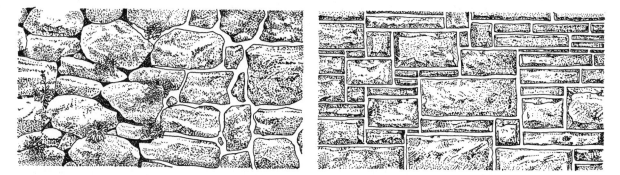

Illustration 10-11. Stone is generally available as rubble (uncut) or as ashlar (cut).

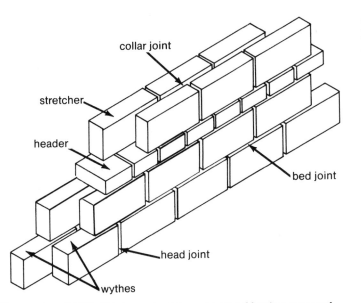

Illustration 10-12. It is important to understand basic terms and bonding directions when planning any concrete masonry work.

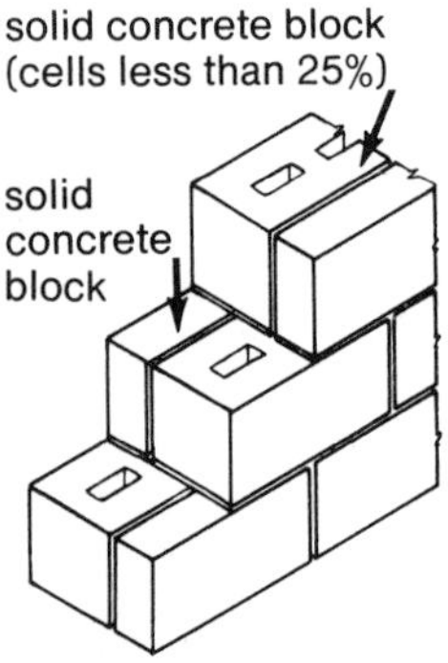

solid concrete block (cells less than 25%)

solid concrete block

Illustration 10-13. A solid masonry wall has less than 25 percent open area in a cross section.

extend 3 to 4 inches into the backing. If the wall does not have headers extending completely through it, headers from the opposite sides overlap 3 to 4 inches. The allowable vertical or horizontal distance between adjacent headers varies from 24 to 36 inches, depending on the local code.

Hollow Masonry Walls

Hollow masonry walls (see illustration 10-14) are built of hollow or combined hollow and solid masonry units laid in a face-shell mortar bedding. All horizontal and vertical edges of the face shells are mortared together.

Hollow masonry walls may be built in any required thickness with single or multiple wythes. Multiwythe hollow masonry walls usually consist of two wythes, facing and backup, and may also be classified as composite walls. The wythes are bonded together and all collar joints are filled with mortar.

Cavity Walls

A cavity wall consists of two walls separated by a continuous air space 2 to 3 inches wide and tied together by rigid metal ties embedded in the mortar joints of both walls (see illustration 10-15). The facing wall usually consists of one wythe of solid or hollow masonry units $3\frac{1}{2}$ to 4 inches thick. The backing may be a single-wythe or multiwythe solid or hollow masonry wall. The thickness of the backing may be equal to or greater than that of the facing, depending on such structural requirements as wall height and the loads to be carried. Usually the cavity wall is designed so that all the vertical loads are carried by the backing; the outer wall serves as a weather-protective facing. Being tied together, both

with mortar or grout. Facing units are usually brick or other solid architectural units that are laid with full head and bed joints. Backup units consist of solid masonry units laid with full head and bed joints.

If units with flanged ends are used, the end cavity must be filled with grout. The collar joints in exterior walls are completely filled by grouting, slushing, or back parging either the facing or backup units and then shoving them into place.

A structural bond between wythes is created by means of masonry headers, unit metal ties, continuous metal ties, or grout. Typical codes require that not less than 4 percent of the area of each wall face be composed of headers. Headers usually consist of facing units laid transversely so that they

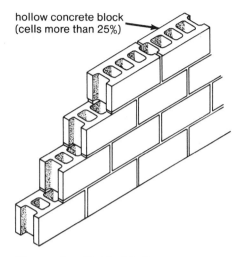

Illustration 10-14. A hollow masonry wall is constructed with units having more than 25 percent open area in a cross section. This type of construction is the most common in residential use.

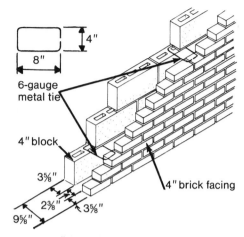

10" Cavity Wall of Block and Brick

Illustration 10-15. Cavity walls have an open area between wythes that permits use of insulation and prevents formation of condensation on interior walls.

walls act to resist the wind, although not necessarily equally.

In areas of severe weather exposure, the wall cavity offers three main advantages.

• It increases the insulating value of the wall and permits use of insulation within the wall.

• It prohibits the passage of water or moisture across the wall.

• It prevents the formation of condensation on interior surfaces; therefore, plaster may be applied directly to the interior masonry, or it may be painted or left unfinished.

Insulation placed within the cavity consists of mats, rigid boards, or non-water-absorbent fill material such as water-repellent vermiculite or silicone-treated perlite. Mats or rigid boards may be fiberglass, foamed glass,

or foamed plastics. A vapor barrier or *damp proofing* is required on the cavity face of the inner wall unless water-proofed insulation is used or the insulating rigid boards are held at least 1 inch away from the exterior wall.

Composite Walls

A composite wall is a multiwythe wall having at least one of the wythes dissimilar to the other(s) with respect to type or grade of masonry unit or mortar, although each wythe contributes to the strength of the wall (see illustration 10-16). A common example is brick or stone bonded to concrete masonry units. Because a composite wall containing different materials has wythes that are bonded together, it is not considered a veneered wall.

In building 8- and 12-inch-thick walls with brick facing and concrete block backup, every seventh course of

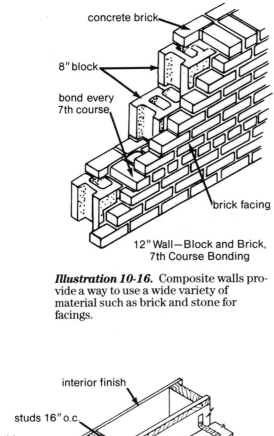

concrete brick

8" block

bond every
7th course

brick facing

12" Wall—Block and Brick,
7th Course Bonding

Illustration 10-16. Composite walls provide a way to use a wide variety of material such as brick and stone for facings.

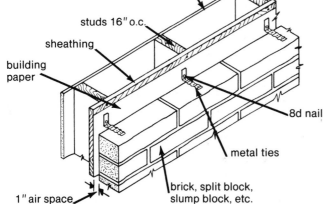

interior finish

studs 16" o.c.

sheathing

building
paper

8d nail

metal ties

1" air space

brick, split block,
slump block, etc.

Illustration 10-17. Masonry veneers are commonly constructed over a wooden frame using metal ties to anchor the veneer to the frame.

brick is a header course. The 12-inch walls can also be bonded every sixth brick course by using concrete block header units. (The masonry headers consist of solid or hollow masonry units lapping over the courses below.)

Veneered Walls

It is common practice in residential construction to use masonry veneer as a non-load-bearing siding or facing material over a wooden frame (see illustration 10-17). Designed to carry its own weight only, veneer is anchored but not bonded to the backing.

The purpose of the veneer is to provide a durable, attractive exterior finish that will prevent entrance of water or moisture. An air space of at least 1 inch is provided between the veneer and the wooden backing to give additional insurance against moisture penetration and heat loss. Flashing and weep holes are provided at the bottom of the air space to eliminate water that may penetrate the veneer.

Metal ties anchoring the veneer to the backing are usually 22-gauge corrugated, galvanized steel strips, $7/8$ inches wide. Building code requirements for spacing of such ties vary widely, but an average value would be 16 inches vertically and 32 inches horizontally.

Veneer may also be anchored to the backing by grouting it to paper-backed, welded-wire fabric attached directly to the wooden studding. The thickness of grout between the backing and the veneer is at least 1 inch. No sheathing is required, although it may be added for stiffness, and the need for flashing and weep holes at the base of the wall is eliminated. This type of construction is commonly called *reinforced masonry veneer.*

Reinforced Masonry Walls

Reinforced masonry walls (see illustration 10-18) are used in cases of high stress concentrations, or in areas of high winds or earthquake probabilities. Embedment of steel in grouted vertical and horizontal cavities gives the wall increased strength. This permits the use of higher design stresses and an increase in the distance between lateral supports. Single or multiple wythes may be used.

Wall Patterns

Exposed masonry is an attractive finish wall material for both exteriors and interiors of homes. Different architectural effects can be created by varying the pattern in which units are laid and by applying different treatments to the mortar joints.

For example, if a long, low look is desired, 2-inch-high units 16 inches long will accentuate the horizontal lines. The opposite effect can be achieved by using other size units evenly placed one atop the other to emphasize the vertical lines (stacked bond pattern). Concrete block can also be laid in staggered (running bond), diagonal, and random patterns to produce almost any result you may be seeking.

In some wall treatments all the joints are accentuated by deep tooling; in others only the horizontal joints are accented. In the latter treatment the vertical joints are tooled, refilled with mortar, and then rubbed flush (after the mortar has partially hardened) to give the joints a texture similar to that of the concrete masonry units. This treatment makes the horizontal joints stand out in relief and is well suited to walls where strong horizontal lines are desired. If an espe-

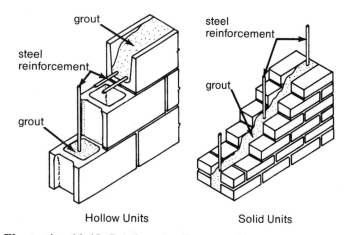

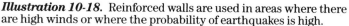

Illustration 10-18. Reinforced walls are used in areas where there are high winds or where the probability of earthquakes is high.

cially massive effect is sought, every second or third horizontal or vertical joint can be accented by having all other joints, both horizontal or vertical, refilled with mortar (after tooling) and rubbed flush.

Numerous wall bond patterns are possible (see illustration 10-19); and variations of these patterns may be created by projecting or depressing the faces of some units from the overall surface of the wall—or by substituting screen block, split block, or customized architectural units with three-dimensional faces.

Illustration 10-19. A rich and varied assortment of wall bond patterns can be created with brick and concrete masonry units.

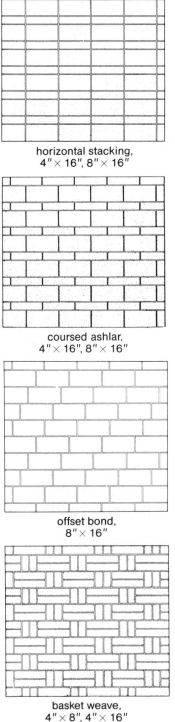

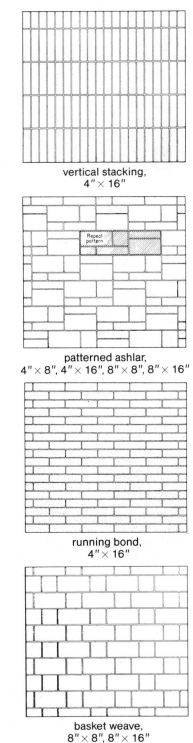

horizontal stacking,
4″ × 16″, 8″ × 16″

vertical stacking,
4″ × 16″

coursed ashlar,
4″ × 16″, 8″ × 16″

patterned ashlar,
4″ × 8″, 4″ × 16″, 8″ × 8″, 8″ × 16″

offset bond,
8″ × 16″

running bond,
4″ × 16″

basket weave,
4″ × 8″, 4″ × 16″

basket weave,
8″ × 8″, 8″ × 16″

Masonry Construction Techniques

Masonry structures have been built for thousands of years and the methods passed on from generation to generation by instruction and example. Because masonry skills are best learned by watching and doing, it's a good idea to look for an opportunity to watch an experienced mason work for a couple of days. If you have access to a video camera, take it with you to record critical procedures so you can review the process at home. If you cannot learn directly from a pro, you can teach yourself; but, you ought to have your work checked by a building inspector or a professional soon after starting to make sure you are on the right track.

Material Requirements

The first step in construction is to estimate the quantity of materials required. Tables 11-1 through 11-3 may be used as guides for estimating the quantities of materials you will need for your project.

Material quantities for single-wythe walls are given in table 11-1. The mortar quantities include the customary allowance for waste that occurs during construction for a variety of reasons. The breakdown of materials in approximately 1 cubic foot of mortar is given in table 11-2. Table 11-3 gives grout quantities for grouted concrete block walls. This table is also useful for estimating grout in reinforced concrete masonry.

Construction Procedures

As we have stated before, masonry is a skill acquired with instruction and practice. If you've never done masonry work before, it's a good idea to begin where less than perfect work won't matter. Mix a small amount of mortar and lay brick and block in the most inconspicuous area of your project. Your skill level will increase rapidly as you practice so don't get frustrated too soon.

149

Table 11-1. Material Quantities for Single-Wythe Concrete Masonry Construction

Nominal Wall Thickness, in.	Nominal Size (width × height × length) of Concrete Masonry Units, in.	Material Quantities for 100-sq. ft. Wall Area		Mortar for 100 Units, cu. ft.*
		Number of Units	Mortar, cu. ft.	
4	4 × 4 × 16	225	13.5	6.0
6	6 × 4 × 16	225	13.5	6.0
8	8 × 4 × 16	225	13.5	6.0
4	4 × 8 × 16	112.5	8.5	7.5
6	6 × 8 × 16	112.5	8.5	7.5
8	8 × 8 × 16	112.5	8.5	7.5
12	12 × 8 × 16	112.5	8.5	7.5

NOTE: Based on 3/8-inch mortar joints.
*With face-shell mortar bedding. Mortar quantities include allowance for waste.

Table 11-2. Sample Quantities of Mortar Materials

Mortar Type	Mix Proportions, Parts by Volume					Material Quantities, cu. ft., for approximately 1 cu. ft. of Mortar			
	Portland Cement	Masonry Cement Type S	Masonry Cement Type N	Hydrated Lime	Sand	Portland Cement	Masonry Cement	Hydrated Lime	Sand
S	1	—	—	0.5	4	0.25	—	0.13	1.0
N	1	—	—	1	6	0.17	—	0.17	1.0
N	—	—	1	—	3	—	0.33	—	1.0
S	—	1	—	—	3	—	0.33	—	1.0
S	1/2	—	1	—	4	0.13	0.25	—	1.0

The instructions that follow deal primarily with concrete block, but the same principles hold for brickwork. Concrete block walls should be built on concrete footings that extend at least 6 inches below the frost line, measure twice the width of the proposed wall, and are as thick as the wall.

Storing Materials

Masonry units should be relatively dry when delivered to the job site. To maintain their dry condition, they should be stockpiled on planks or other supports free from contact with the ground and then covered with canvas or polyethylene tarpaulins.

Care must be taken to keep concrete masonry units dry because moisture in the units affects bond. Clay brick requires wetting before use.

Table 11-3. **Volume of Grout in Grouted Concrete Block Walls**

Wall Thickness, in.	Spacing of Grouted Cores, in.	Grout, cu. yd., for 100-sq. ft. Wall Area*	Wall Area, sq. ft., for 1 cu. yd. of Grout*
6	all cores grouted	0.79	126
	16	0.40	250
	24	0.28	357
	32	0.22	450
	40	0.19	526
	48	0.17	588
8	all cores grouted	1.26	79
	16	0.74	135
	24	0.58	173
	32	0.49	204
	40	0.44	228
	48	0.39	257
12	all cores grouted	1.99	50
	16	1.18	85
	24	0.91	110
	32	0.76	132
	40	0.70	143
	48	0.64	156

*A 3 percent allowance has been included for waste and job conditions. All quantities include grout for intermediate and top bond beams in addition to grout for cores.

Concrete masonry units should never be wetted immediately before and during placement, and the top of a masonry structure should be covered with tarpaulins or boards to prevent rain or snow from entering unit cores during construction. When moist concrete units are placed in a wall, they will shrink with the loss of moisture. If this shrinkage is restrained, as it usually is, stresses develop that may cause cracks in the walls. Hence, it is important that the units be kept dry before use. Clay bricks do not react in this manner. Rather than being kept dry, brick should actually be wetted several hours before use.

Sometimes it may be advisable to dry concrete masonry units below the moisture content usually specified for the locality—for example, where walls will be exposed to relatively high temperature and low humidity in interiors of heated buildings. In such cases it is advisable that, before placement, the units be dried to approximately the average air-dry condition to which the finished construction will be exposed in service.

Damp concrete masonry units can be stacked to facilitate drying and then be artificially dried by blowing heated air through the cores and the spaces between the stacked units. An inexpensive and efficient drying device consists of a combination oil- or gas-burning heater and fan. This method of drying works equally well indoors or outdoors and can readily be used at the work site.

Mortaring Joints

Two types of mortar bedding are used with concrete masonry: full mortar bedding and face-shell mortar bedding (see illustration 11-1). In full mortar bedding, the unit webs as well as face shells are bedded in mortar. Full bedding is used for laying the first or starting course of block on a footing

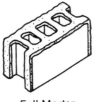

Full Mortar
Bedding

Face-shell
Mortar Bedding

Illustration 11-1. There are two types of
mortar bedding depending upon the load-
bearing capacity desired.

As you become more skilled at laying
block, you'll be able to butter and spread
bed mortar for three blocks at a time
allowing you to lay block in rapid
succession. This is convenient because
three blocks plus mortar joints equal 4
feet—the length of a mason's level.

or foundation wall as well as for laying
solid units such as concrete brick and
solid block. It is also commonly used
when building concrete masonry
columns, piers, and pilasters that will
carry heavy loads. Where some vertical cores are to be solidly grouted,
such as in reinforced masonry, the webs
around each grouted core are fully
mortared. For all other concrete masonry work with hollow units, it is common practice to use only face-shell
bedding. Also, the head (vertical) joints
of block having plain ends are mortared
only opposite the face shells.

Block   For bed (horizontal) joints,
all concrete block should be laid with
the thicker part of the face shell up.
This provides a larger mortar-bedding
area and makes the block easier to lift.
For head joints, mortar is applied only
on the face-shell ends of block. Some
masons butter (mortar) the vertical
ends of the block previously placed;
others set the block on one end and
butter the other end before laying the
block. Time can be saved by placing
three or four blocks on end and then
buttering their vertical edges in one
operation. If you want, you can butter
both the block already laid and the
block to be laid; such application of
mortar ensures well-filled head joints.

Regardless of the method used to
apply mortar to the vertical edges, each

A well-filled head joint results from
mortaring both units.

block is brought over its final position
and pushed downward into the mortar bed and sideways against the
previously laid block so that mortar
oozes out of the head and bed joints
on both sides of the face shell. This
indicates that the joints are well filled.
CAUTION: Mortar should not be
spread too far ahead of the actual laying
of units or it will tend to stiffen and
lose its plasticity, thereby resulting in
poor bond. In hot, dry weather it may
be necessary to spread only enough
mortar for each block as it is laid.

As each block is laid, excess mor-

Each unit is pushed downwards and sideways so that mortar oozes out of the head and bed joints.

Brick is pressed into place so that mortar oozes out of the head and bed joints.

For laying brick, spread mortar uniformly on the bed joints.

tar extruding from the joints is cut off with the trowel and thrown back on the mortarboard for reuse. You may apply the extruded mortar to the vertical face shells of the block just laid. If there has been any delay long enough for the extruded mortar to stiffen on the block before it is cut off, it should be reworked on the mortarboard before

reuse. Mortar droppings picked up off the scaffolding, floor, or ground should not be reused.

Brick For concrete and clay brick, mortar should be spread uniformly thick for the bed joints and furrowed only slightly if at all. (**NOTE:** Some building codes prohibit furrowing on bed joints.) The weight of the brick and the courses above help compact the mortar and ensure watertight bed joints.

In brick construction, special care should be taken in filling the head joints because they are more vulnerable to water penetration than the bed joints. If head joints are not completely filled with mortar, some voids and channels may permit water to penetrate to the inside of the wall. Plenty of mortar should be troweled on the end of the brick to be placed so that, when the brick is shoved into place, the mortar will ooze out at the top and around the sides of the head joint, indicating the joint is completely filled. Dabs of mortar spotted on both corners of the brick do not completely fill the head joints, and *slushing* (attempting to fill the joints from above after the brick is placed) cannot be relied on to fill all voids left in the head joints.

Begin a masonry wall by stringing out the units without mortar along a chalk line.

For the first course, a full mortar bed is placed on a clean, properly cured footing. Place no mortar on areas under cores that will be grouted.

Laying Up a Wall: First Course

A concrete block wall is laid on a poured concrete footing after the footing has properly cured. Begin construction by stringing out the masonry units for the first course without mortar in order to check the wall layout. A chalk line is sometimes used to mark the foundation and help align the block accurately. If the footing is badly out of level, the entire first course should be laid before you can begin work on other courses.

Before any units are laid, the top surface of the concrete foundation should be clean. Remove laitance and expose aggregate by sandblasting, chipping, or scarifying if necessary to ensure a good bond of the masonry to the foundation. Then spread a full bed of mortar and furrow it with a trowel to ensure plenty of mortar along the bottom edges of the block for the first course. If the wall is to be grouted, the mortar bedding for the first course should not fill the area under the block cores—grout should come into direct contact with the foundation.

The corner block should be laid first and carefully positioned. After three or four blocks have been laid, use a mason's level as a straightedge to assure correct alignment of the block. Carefully check these units with the

Corners are laid first with great care. Use a level as a straightedge to assure correct alignment of the units.

Units are leveled by tapping with the trowel handle.

The trowel handle is also used to make the block plumb (vertically straight).

Check the corner for plumb.

Lay up corners four to six units high and check each course for alignment.

Check each corner for level.

Check each block as you lay it to see that it is in the same plane.

level. Bring them to proper grade and plumb by tapping with the trowel handle. The entire first course of a concrete masonry wall should be laid with such care, making sure each unit is properly aligned, leveled, and plumbed. This will assist you in laying succeeding courses. Any error at this stage in the first course means continuing trouble in laying up a straight, true wall.

Corners The corners of the wall are built up higher, usually four or five courses higher, than the course being laid at the center of the wall. As each course is laid at a corner, check it with a level for alignment, for level, and for plumb. In addition, check each block carefully with a level or straightedge to make certain that the faces of all the block are in the same plane.

Use the level to make a diagonal check of the horizontal spacing of units.

Use a story pole to check the spacing from course to course.

When filling in the wall between the corners, a mason's line should be stretched from corner to corner for each course.

Other precautions are necessary at the corners to ensure true, straight walls. An accurate method for finding the top of the masonry in each course is provided by the use of a simple story pole made from a 1 × 2-inch wooden strip with markings 8 inches apart. More elaborate metal story poles are commercially available. Also, because each block in a course overlaps the one below by a half unit, you can easily check the horizontal spacing of the units by placing your level diagonally across their corners.

Between Corners After the corners at each end of a wall have been laid up, a mason's line (string line) is stretched tightly from corner to corner for each course, and the top outside edge of each block is laid to this line. The line is moved up as each course is laid.

The manner of handling and gripping a masonry unit is important, and the most practical way for each individual is determined through practice. Generally, you tip a block slightly toward yourself so that you can see the upper edge of the course below and thus place the lower edge of the block directly over it. By rolling the block slightly to a vertical position and shoving it against the previously laid unit, you can lay the block to the mason's line with minimum adjustment. This speeds the work and reduces the possibility of breaking mortar bond by not moving the unit excessively after it has been pressed into the mortar. Light tapping with the trowel handle should be the only adjustment necessary to level and align the unit to the mason's line. The use of the mason's level between corners is limited to checking the face of each unit to keep it aligned in a true plane with the face of the wall.

CAUTION: All adjustments to final position must be made while the mortar is soft or plastic. Any adjustments made after the mortar has stiffened or even partially stiffened will break the mortar bond and cause cracks between the masonry unit and the mortar. This will allow penetration of water. Any unit disturbed after the mortar has stiffened should be removed and relaid with fresh mortar. Realignment of a unit should not be attempted after a higher course has been laid.

Care must be taken to keep the wall surface clean during construction. In removing excess mortar that has oozed out at the joints, you must avoid smearing soft mortar onto the face of the unit, especially if the wall is to be left exposed or painted. Numerous embedded mortar smears will detract from the neat appearance of the finished wall. They can never be removed and paint cannot be depended on to hide them.

Any mortar droppings that do stick to the wall should be almost dry before they are removed with a trowel. Then when dry and hard, most of the remaining mortar can be removed by rubbing with a small piece of concrete masonry and by brushing.

Closure Unit Before the *closure* unit is laid in a course, the length of the opening is checked in order to avoid joints that are too tight or too wide. If

The top outside edge for each course is laid to the line. The manner of handling or gripping the blocks is important. Practice determines the best way for you.

Light tapping brings the block into position with the string line.

necessary, the closure unit is accurately measured, sawed, and dressed for a proper fit in the opening. If sawing is required, a masonry blade on a hand-held circular saw will do the job. Clay bricks may be cut using a hammer and brick set. To cut a brick, place

To prevent smears, mortar droppings should be almost dry when cut off the wall with a trowel. Most of the dry mortar remaining can be rubbed off with a scrap piece of concrete masonry.

Closing a course begins by buttering all edges of the opening.

it on the ground, score it with the brick set and hammer, then cut it with a sharp blow to the brick set.

When installing the closure unit, you butter all edges of the opening and of the closure unit before you carefully lower the unit into place. If any of the mortar falls out, leaving an open joint, remove the closure unit, apply fresh mortar, and repeat the operation. Closure unit locations are staggered throughout the length of the wall.

Building Reinforced Walls

For reinforced masonry wall construction the procedures used in laying masonry units, placing reinforcing bars, and pouring grout vary with the size of the job. This section covers only the general requirements of common procedures.

Solid or hollow concrete masonry units should be laid so that their alignment forms an unobstructed, continuous series of vertical cores within the wall framework. Spaces in which reinforcement will be placed should be at least 2 inches wide. No grout space should be less than 3/4 inch or more than 6 inches wide; if the grout space is wider than 6 inches, the wall

The edges of the closure block are buttered, then it is carefully lowered into place.

section should be designed as a reinforced concrete member.

Two-core, plain-ended units are preferable to three-core units because the larger cores allow easier placement of reinforcing bars and grout. Also, these units are more easily aligned for their cores to form continuous, vertical spaces.

The mortar bed under the first course of block should not fill the core area because the grout must come into direct contact with the foundation. All head and bed joints should be filled solidly with mortar for the thickness of the face shell. With plain-ended units, however, it is not necessary to fill the head joint across the full unit width. Also, when the wall is to be grouted intermittently (for reinforcement 16, 24, 32, or 48 inches on center), only the webs at the extremity of those cores containing grout are mortared. When the wall is to be solidly grouted, none of the cross webs need be mortared because it is desirable for the grout to flow laterally and form the bed joints at all web openings.

Mortar protrusions that cause bridging and thus restrict the flow of grout require an excessive amount of vibration to assure complete filling of the grout space. Hence, care is necessary that mortar projecting more than $3/8$ inches into the grout space be removed and that excess mortar does not extrude and fall into the grout space. You can prevent mortar from extruding into the grout space by placing the mortar no closer than $1/4$ to $1/2$ inch from the edge of the grout space and troweling the mortar bed upward and outward, away from the edge, thus forming a bevel.

Vertical reinforcement may be erected before or after the masonry units are laid. When the reinforcing bars are placed before the units, the use of two-core, open-ended, A- or H-shaped units becomes desirable in order for the units to be threaded around the reinforcing steel. When the bars are placed after the units, adequate positioning devices are required to prevent displacement during grouting. Both vertical and horizontal reinforcement should be accurately positioned and rigidly secured at intervals by wire ties or spacing devices (see illustration 11-2).

Horizontal reinforcement is placed as the wall rises. The reinforcing bars

are positioned in bond-beam, lintel, or channel units, which are then solidly grouted (see illustration 11-3). Where the wall itself is not to be solidly grouted and cored bond-beam units are used, the grout may be contained over open cores by placing expanded *metal lath* in the horizontal bed joint before the mortar bed is spread for the bond-beam units. Paper or wood should not be used as a grout barrier because of lack of fire resistance.

To ensure solid grouting of bond beams, it may be necessary to fill those portions of the bond beams between the vertically grouted cores as the bond-beam courses are laid, especially if the spacing of vertically grouted cores is greater than 4 feet. Otherwise, the grout may not flow far enough horizontally from the cores being grouted to completely fill the bond beams.

A concrete masonry wall should be grouted as soon as possible to reduce shrinkage cracking of the joints. However, placing grout before the mortar has been allowed to cure and gain strength may cause shifting or blow-out of the masonry units during the grouting operations. Therefore, to fill large cavities of masonry sections (made up of two or more units and containing vertical joints, such as pilaster sections), grout should be poured after the mortar in the entire height of the masonry has been cured a minimum of 3 days during normal weather or 5 days during cold weather. When filling hollow-unit masonry walls, it is unnecessary to cure mortar for longer than 24 hours before grouting.

Grouting

In grouting of a single-wythe wall, the wall is built to a height not exceeding 5 feet before grout is poured into the cores. This operation is repeated by alternately laying units and grouting at successive heights not exceeding 5

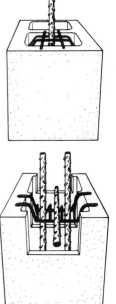

Illustration 11-2. Reinforcing bar spacers are used to hold vertical and horizontal reinforcement in place until grout is poured.

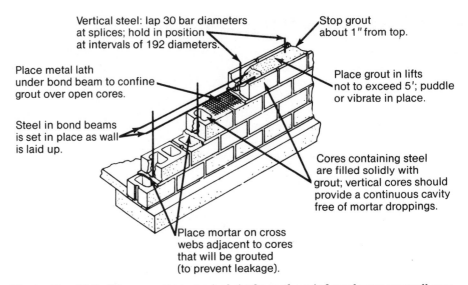

Vertical steel: lap 30 bar diameters at splices; hold in position at intervals of 192 diameters.

Stop grout about 1″ from top.

Place metal lath under bond beam to confine grout over open cores.

Place grout in lifts not to exceed 5′; puddle or vibrate in place.

Steel in bond beams is set in place as wall is laid up.

Cores containing steel are filled solidly with grout; vertical cores should provide a continuous cavity free of mortar droppings.

Place mortar on cross webs adjacent to cores that will be grouted (to prevent leakage).

Illustration 11-3. When grouting a typical single-wythe reinforced masonry wall, you can use a combination of reinforcing bars and lath to provide reinforcement to just those spaces that require it.

feet. Each of these grouted layers is referred to as a *lift*.

Vertical cores to be filled should have an unobstructed alignment, with a minimum dimension of 2 inches and a minimum area of 8 square inches. Also, the vertical reinforcing bars may be relatively short in length because they only need to extend above the top of the lift a distance equal to 30 bar diameters for sufficient overlap with the reinforcing bars in the next lift. As an alternate choice, vertical steel may extend to full wall height for one story construction. However, because the long lengths of steel require the use of open-ended units, you may want to lap the steel just above each 5-foot lift.

Grout is handled from the mixer to the point of deposit in the grout space as rapidly as practical. On small projects, the grout is poured with buckets having spouts or funnels to confine the grout and prevent splashing or spilling onto the face or top surface of the masonry. Grouting should be done from the inside face of the wall if the outside will be exposed; dried grout can deface the exposed surface of a wall and be detrimental to the mortar bond of the next masonry course.

Whenever work is stopped for 1 hour or longer, a horizontal construction joint should be made by stopping the grout pour about 1 inch below the top of the masonry unit to form a key with the next lift.

During placement, grout should be rodded (usually with a 1 × 2-inch wooden stick) to ensure complete filling of the grout space and solid embedment of the reinforcement. It takes very little effort to do this consolidation job properly because of the fluid consistency of the grout. When high-absorption masonry units are used, it may be necessary to rerod the grout 15 to 20 minutes after placement to overcome any separations (of the grout from the reinforcing steel) and voids caused by settlement of the grout and absorption of water into the surrounding masonry.

Anchor Bolts

In construction of concrete masonry with wooden framing, wooden plates are fastened to the tops of concrete block walls with $\frac{1}{2} \times 18$-inch anchor bolts spaced not more than 4 feet apart. The bolts are placed in the cores of the top two courses of block and the cores then filled with concrete, grout, or mortar. Pieces of metal lath are placed in the second horizontal mortar joint from the top of the wall and under the cores that will be filled to hold the concrete grout or mortar filling in place. The threaded end of the bolt should extend about 3 inches above the top of the wall; and when the filling has hardened, a wooden plate can be securely fastened to the wall.

Tooling Mortar Joints

Weathertight joints and the neat appearance of concrete masonry walls are dependent on proper *tooling;* that

Anchor bolts grouted into the cores of concrete masonry units provide the necessary means to fasten wooden framing to a foundation.

is, compressing and shaping the mortar face of a joint with a special tool slightly larger than the joint. After a wall section has been laid and the mortar has become thumbprint hard (when a clear thumbprint can be impressed and the cement paste does not adhere to the thumb), the mortar joints are usually considered ready for tooling. On hardening, mortar has a tendency to shrink slightly and pull away from the edges of the masonry units. Proper use of a jointing tool restores the intimate contact between the mortar and the units and helps to make weathertight joints by sealing any cracks left between the mortar and the units when they were laid. Proper tooling also produces uniform joints with sharp, clean lines.

Horizontal joints should be tooled before vertical joints. A jointer for tooling horizontal joints should be at least 22 inches long, preferably longer, and upturned on at least one end to prevent gouging, while a jointer for vertical joints is small and S-shaped. Plexiglass jointers are available to avoid staining white or light-colored mortar joints. After the joints have been tooled, any mortar burrs should be trimmed off flush with the face of the wall by using a trowel or by rubbing with a burlap bag, brush, or carpet.

There are several principal types of mortar joints used in masonry (see illustration 7-2). The concave, V-shaped, raked, and beaded types need special jointing tools, whereas the flush, struck, and weathered types are finished with a trowel. The *extruded* (also called skintled or weeping) type is made by using extra mortar so it can be squeezed out or extruded as the units are laid; it is not trimmed off but left to harden. Extruded joints are not recommended for walls subject to heavy rains, high winds, or freezing temperatures.

When the mortar is thumbprint hard, tooling is begun by running a jointer (or sled runner) along the horizontal joints.

After the horizontal joints, the vertical joints are tooled using a small S-shaped jointer.

V-shaped joints are usually narrow in appearance and have sharp shadow lines. *Concave* joints have less pronounced shadows. Both types are tooled and very effective in resisting rain penetration. They are recommended for exterior weathertight walls, as is the weathered type.

Flush joints are simple to make because the excess mortar is simply trimmed off with a trowel (striking downward rather than upward) and the flush face rubbed with a carpet-covered wooden float. The mortar is not compacted by tooling, and small hairline cracks produced when the mortar is pulled away from the units by the trimming action may permit infiltration of water into the wall. Flush joints are used in walls that will be plastered.

Raked joints are made with a special tool, a joint raker or skate, to remove the mortar to a certain depth, which should not be more than $1/2$ inch. These joints produce dark shadows that accent the masonry pattern. Because their ledges may hold rain, snow, or ice that may affect the watertightness of the wall, they are best suited to dry climates or interior use.

Beaded joints are basically extruded joints that are tooled with a

Mortar burrs are trimmed off after the joints are tooled and then dressed with a burlap bag or a brush.

special bead-forming jointer. The beads protruding on the wall surface present strong shadow lines, but special care is required to obtain a straight appearance.

The *struck* and *weathered* types of joints are generally used to emphasize horizontal lines. Struck joints are easy to make with a trowel, especially if you work from the inside of the wall. However, their small ledges do not shed water readily, making them unsuitable for use in areas where heavy rains, driving winds, or freezing temperatures are likely to occur. On the other hand, weathered joints (a type recommended for weathertight walls) shed water easily but require careful finishing; that is, they must be worked with a trowel from below.

The overall appearance of a masonry wall depends not only on the joint treatment but also on the color uniformity of the joints. Although mortar shade is influenced to some degree by the moisture condition of the units and by the atmospheric conditions, it depends mainly on the uniformity of the mortar mix and the time of joint tooling. The amount of water used in mixing colored mortar greatly influences the shade and thus requires accurate control. Retempering of colored mortar should be avoided, and any mortar that has become too stiff for use should be discarded. A darker color results if the mortar is tooled when relatively hard rather than reasonably plastic, but some masons consider mortar ready for tooling only when thumbprint hard. Uniform time of tooling is important for obtaining uniformly colored joints.

Patching and Pointing

In spite of good workmanship, joint patching or *pointing* may sometimes be necessary. Mortar in a head joint may have fallen out while the units were being placed, or a mortar crack may have formed while the units were being aligned. Furthermore, there may not have been enough mortar in a joint to fill the space left by a broken corner or edge. Sufficient additional mortar should be forced into such spots to completely fill the joints.

Patching or pointing is done preferably while the mortar in the joint is still fresh and plastic. If the back of the face shell can be reached when forcing additional mortar into the joint, provide a backstop, such as the handle of a hammer. Any depressions and holes made by nails or line pins are filled with fresh mortar before tooling.

When patching or pointing must be done after the mortar has hardened, the joint is chiseled out to a depth of about $1/2$ inch, thoroughly wetted, and repointed with fresh mortar.

Hot-Weather Construction

Hot weather poses some special problems for concrete masonry construction. These arise, in general, from higher temperatures of materials and equipment and more rapid evaporation of the water required for cement hydration and curing. Other factors contributing to the problems include wind velocity, relative humidity, and sunshine.

As the temperature of mortar increases, there are several accompanying changes in its physical properties.

• Workability is lessened; that is, for a given workability, more water is required.

• A given amount of air-entraining agent will yield less entrained air.

• Initial and final set will occur earlier while evaporation will generally be faster.

• Depending on the surface characteristics, temperature, and moisture content of the concrete masonry units, their suction of moisture from the mortar will be faster.

The result of these changes is that mortar will rapidly lose water needed for hydration. Despite its higher initial water content, mortar will be somewhat more difficult to place and the time available for its use will be shorter.

Early surface drying of mortar joints is particularly harmful. Evaporation removes moisture more rapidly from the outer surface of mortar joints, but the inner parts retain moisture longer and so develop greater strength. Weak mortar on the surface reduces the strength of the wall under wind and other horizontal loads.

Selection and Storage of Materials

During hot weather there is a temptation to reduce the amount of cementitious material in the mortar mixture in order to lessen the heat of hydration released at early stages. Actually, the better solution is to increase the amount of cementitious material. This will accelerate rather than retard the mortar's gain in early strength and thus secure maximum possible hydration before water is lost by evaporation.

Mortar materials stored in the sun can become hot enough to significantly affect the temperature of the mortar mixture itself. Covering or shading such materials from the sun can be helpful. For example, sand delivered to the job site normally contains free moisture ranging from 4 to 8 percent, which is sufficient to ensure that a covered or shaded stockpile of sand remains reasonably cool. Of course, if the moisture content drops much below this level, the stockpile should be sprinkled to increase evaporative cooling. When evaporating, 1 gallon of water will cool 1 cubic yard of sand 20 degrees at the surface of a stockpile.

In hot weather, the main objective is to see that all of the materials of masonry construction are placed without having acquired excess heat. Heat should be minimized in masonry units by storing them in a cool place and the mortar mixture should be relatively cool. The most effective way of cooling mortar during mixing is to use cool water. Immediately after mortar has been mixed, it begins to rise in temperature and must be protected from further heat gain during construction.

Other Construction Practices

Attention should be given to cooling metal equipment with which the masonry materials, particularly mortar, come into contact. Relatively cool mortar can heat rapidly when transported in a metal wheelbarrow or other container that has been exposed for hours to the sun's rays. Metal mortarboards can become quite hot and wooden ones can become very absorptive in hot weather. Flushing them with water immediately before use and/or working under sunshades can lessen such difficulties.

Because wind and low relative humidity cause increased evaporation, the use of wind screens and fog (water) sprays can effectively reduce the severe effects of hot, dry, windy weather. Also, covering walls immediately after construction will effectively slow the rate of loss of water from masonry. Damp curing is very effective, particularly in development of tensile bond. If the wall will be subjected to flexure, consideration should be given to damp curing.

In areas where high air temperatures are common, avoid construction during the hot, midday periods. Construction during evening hours or during the early morning hours can avoid many hot-weather problems.

Cold-Weather Construction

When the temperature falls below 40°F, your productivity and workmanship as well as performance of materials may be lowered. During cold weather you are not only concerned with normal construction tasks but also with personal comfort, additional materials preparation and handling, and protection of the structure. As temperatures continue to drop, these extra activities consume more time.

Masonry Performance at Low Temperatures

Immediately after concrete masonry units are laid during cold weather, several factors come into play. The absorptive masonry units tend to withdraw water from mortar; but mortar, having the property of retentivity, tends to retain water. The surrounding air may chill masonry as well as withdraw water through evaporation. Also, if the masonry units are cold when laid, they will drain heat from mortar. Any combination of these factors influences strength development.

As the temperature falls below about 40°F, mortar ingredients become colder and the heat-liberating reaction between portland cement and water is substantially reduced. Hydration and strength development are minimal at temperatures below freezing. However, construction may proceed if the mortar ingredients are heated. The masonry units should also be heated and the structure maintained above freezing during the early hours after construction.

Mortars mixed with cold but unfrozen materials possess plastic properties quite different than those at normal temperatures. The water requirements for a given consistency decrease as the temperature falls; more air is entrained with a given amount of air-entraining agent, and initial and final set are delayed. Also, with lower temperature, the strength gain of mortar is less, although final strength may be as high or higher than that of mortar used and cured at more normal temperatures.

Heated mortar materials produce mortars with performance characteristics identical to those in the normal-temperature range, and thus heating is desirable for cold-weather masonry construction. Mortars mixed to a particular temperature and subjected to a lower level lose heat until they reach the ambient temperature. If the ambient temperature is below freezing whenever the mortar temperature reaches 32°F, the mortar temperature remains constant until all water in the mortar is frozen. Afterward, the mortar temperature continues to descend until it reaches the level of the ambient temperature.

The rate at which masonry freezes is influenced by the severity of air temperature and wind, the temperature and properties of masonry units, and the temperature of mortar. When fresh mortar freezes, its performance characteristics are affected by many factors: for example, water content, age at freezing, and strength development prior to freezing. Frozen mortar takes on all the outward appearances of hardened mortar, as evidenced by its ability to support loads as well as its ability to bond to surfaces.

Mortar possessing a high water content expands when it freezes; and, the higher the water content, the greater the expansion. The expansive forces will not be disruptive if moisture in the freezing mortar is below 6 percent. Therefore, every effort should be made to achieve mortar with low water content. Dry masonry units and protective coverings should be used.

Mortar that is allowed to freeze gains very little strength and some permanent damage is certain to occur. If the mortar has been frozen just once at an early age, it may be restored to nearly normal strength by providing favorable curing conditions. However, such mortar is neither as resistant to weathering nor as watertight as mortar that has never been frozen.

Material Selection

Cold-weather concrete masonry construction generally requires only a few changes in the mortar mixture. Concrete masonry units used during normal temperatures may be successfully used during cold weather. Under the prevailing recommendations for cold weather masonry construction (see table 11-4), the masonry units will generally lower the moisture within the mortar to below 6 percent; thus, any subsequent accidental freezing will not be disruptive.

At low temperatures mortar performance can be improved with an early strength gain by use of Type III high-early-strength cement. Also, for mortar made with lime, the dry, hydrated form of lime is preferred because it requires less water.

Admixtures often considered for inclusion in mortar are antifreezes, accelerators, corrosion-inhibitors, air-entraining agents, and color pigments. Those used with proven success in cold weather are the accelerators and air-entraining agents.

Table 11-4. **Recommendations for Cold-Weather Masonry Construction**

Air Temperature, °F	Construction Requirements	
	Heating of Materials	**Protection**
Above 40	normal masonry procedures	cover walls with plastic or canvas at end of workday to prevent water entering masonry
Below 40	heat mixing water; maintain mortar temperatures between 40° and 120°F until placed	cover walls and materials to prevent wetting and freezing; covers should be plastic or canvas
Below 32	in addition to the above, heat the sand; frozen sand and frozen wet masonry units must be thawed	with wind velocities over 15 mph, provide windbreaks during the workday and cover walls and materials at the end of the workday to prevent wetting and freezing; maintain masonry above 32°F by using auxiliary heat or insulated blankets for 16 hours after laying masonry units
Below 20	in addition to the above, dry masonry units must be heated to 20°F	provide enclosures and supply sufficient heat to maintain masonry enclosure above 32°F for 24 hours after laying masonry units

Certain admixtures for mortar are misunderstood in that they accelerate strength gain rather than lower the freezing point. So-called antifreeze admixtures, including several types of alcohol, must be used in great quantities to significantly lower the freezing point of mortar, but the compressive and bond strengths of masonry are also lowered. Therefore, antifreeze compounds are not recommended for cold-weather masonry construction.

The primary interest in accelerators is to increase rates of early-age strength development—that is, to hasten hydration of portland cement in mortar. Accelerators include calcium chloride, soluble carbonates, silicates and fluosilicates, calcium aluminate, and organic compounds such as triethanolamine. Aluminous cements and finely ground hydrated cements have also been advocated for acceleration. These materials may be available at your local building supply company where experienced salespersons can help you with the best materials for your local conditions.

The most commonly used accelerator in concrete is calcium chloride. However, its use in mortar is controversial because of possible adverse side effects, such as increased shrinkage, efflorescence, and corrosion of embedded metal. Because calcium chloride may produce corrosion failure, it should not be permitted in mortar for concrete masonry containing metal ties, anchors, or joint reinforcement. If these factors are not involved, it is recommended that the amount of calcium chloride used should not exceed 2 percent by weight of portland cement or 1 percent by weight of masonry cement, added in solution form with the mix water.

Air-entraining agents may be added if you are using a mixer to increase mortar workability and freeze-thaw durability at later ages, but their effects on mortars subjected to early freezing have not been established. In practice, however, air-entrained masonry cement performs satisfactorily in mortars used during winter construction.

Some color pigments contain dispersing agents to speed the distribution of color throughout the mortar mixture. The dispersing agents may have a retarding effect on the hydration of portland cement, and this *retardation* is particularly undesirable in cold-weather masonry construction. In addition, the masonry may have a greater tendency to effloresce.

Storage and Heating of Materials

At delivery time all masonry materials should be adequately protected against any adverse exposure conditions. When the temperature is below 40°F, safe storage can be accomplished by covering all masonry materials. Bagged materials and masonry units should be securely wrapped with canvas or building plastic and stored above the reach of moisture migrating from the ground. The masonry sand should be covered to keep out snow and ice buildup before the sand pile is heated.

The most important consideration in heating materials is that sufficient heat be provided to ensure cement hydration in mortar. After all materials are combined, the mortar temperature should be within the range of 40° to 120°F. If the air temperature is falling, a minimum mortar temperature of 70°F becomes worthwhile. Mortar temperature in excess of 120°F may cause excessively fast hardening with a resultant loss of compressive and bond strength.

Water

When the air temperature drops, water is generally the first material heated for two reasons: (1) it is the easiest material to heat, and (2) it can store the most heat, pound for pound, of any of the materials in mortar. Recommendations vary as to the highest temperature to which water should be heated. Some specifiers put a maximum of 180°F on heating water because there is a danger of flash set if significantly hotter water comes into contact with cement. Combining sand and water in the mixer first, before adding the cement, will lower the temperature and avoid this difficulty. With this precaution and the use of aggregates that are cold enough, even boiling water may be used successfully.

Sand

When the air temperature is below 32°F, sand should be heated so that all frozen lumps are thawed. Generally the temperature of the sand is raised to 45° to 50°F. However, if the need exists and facilities are available, there is no objection to raising the sand temperature much higher: 150°F is a reasonable upper limit.

Sand should not be heated to a temperature that would cause decomposition or scorching. A practical method is to limit the sand temperature by touch. The stockpile must be mixed periodically to ensure uniform heating. A commonly used method of heating sand is to pile it over a metal pipe containing a fire.

Masonry Units

Masonry units can be heated by stacking them around oil-burning salamanders or oil hot-air heaters with built-in blowers. During very cold weather, frozen walls must be heated before grout is poured into cores or cavities. It is recommended that when the air temperature is below 40°F, or it has been below 32°F during the previous 2 hours, the air temperature in the bottom of the grout space should be raised above 32°F before grouting. Heated enclosures may be used for this purpose.

Other Cold-Weather Construction Practices

One of the most important practices of masonry work in subfreezing temperatures is to have the masonry units delivered dry and kept dry until they are laid in the wall. In addition, they should be laid only on a sound, unfrozen surface and never on a snow- or ice-covered base or bed; not only is there danger of movement when the base thaws, but no bond will develop. It is also considered good practice to heat the surface of any existing masonry units to be added. The heat should be sustained long enough to thaw the surface thoroughly.

During cold weather the mortar should be mixed in smaller quantities than usual to avoid excessive cooling before use. To avoid premature cooling of heated masonry units, only those units that will be used immediately should be removed from the heat source.

Regardless of temperature, concrete masonry units should never be wetted before being laid in a wall. In cold weather, wetting masonry units adds to the possibility of freezing, increasing shrinkage, and defeating the goal of drawing off water from the mortar to a level below 6 percent (to prevent mortar expansion upon freezing).

Tooling time during cold weather is less critical than at normal temperatures. Hence, you may lay more masonry units before tooling. In all instances, however, the joint should be tooled before mortar hardens. Also remember, premature tooling will cause lighter joints. So leave enough time after the final units are laid at the end of the day before tooling.

On completion of each section or at the end of each workday, measures should be taken to protect new masonry construction from the weather. The top of the masonry structure must be protected from rain or snow by plastic or canvas tarpaulins extending at least 2 feet down all sides of the structure.

Cold-Weather Safety

Safety precautions require added emphasis during cold-weather construction for several reasons. You may tend to be distracted because of lack of comfort and the development extraordinary construction problems due to low temperature. You also must deal with hazards such as uncertain footing on ice and snow and clumsiness caused by protective clothing. The use of heaters and flammable materials adds the possibility of fire and asphyxiation.

CHAPTER 12

Finishes for Concrete Masonry

A wide variety of applied finish is possible with concrete masonry construction. The finish to use in any particular case will be governed by the type of structure in which the walls will be used, the climatic conditions to which they will be exposed, and the architectural effects desired.

Paints

The main purposes of painting concrete masonry walls are to add a fresh appearance and color and to alter the surface texture and pattern. Additional purposes are to reduce the passage of sound through the wall and to bar the passage of moisture. The paint selected should not only achieve these goals but also attach itself closely to the underlying surface, retain its appearance during a long life, and be economical.

Some paints breathe; that is, they are water-vapor permeable. They allow water vapor but not liquid water to pass through. Others are nonbreathing, or impermeable to both liquid water and water vapor. Impermeable paints should be applied to the side of a wall where moisture enters, the inner surface. Permeable paints should be applied to the surface where moisture exits, the exterior or outer surface, so that moisture trapped inside a wall can escape outdoors. If the surface from which moisture is attempting to leave is coated with an impermeable paint, blistering will occur and the paint will peel.

There are several other items that must be considered when selecting paint for a concrete masonry surface. Among them are whether or not the paint will be damaged by the presence of alkalies in the concrete, and whether the surface texture requires alteration before application of the paint.

Commonly Used Paints

Many different paint products have been marketed for use on concrete masonry walls with varying degrees of success. In addition, urethanes, polyesters, and epoxies are also used successfully. We will now describe the basic constituents and pertinent characteristics of the more common and successful masonry paints.

Portland Cement Paints   Of the various types of paint used on concrete masonry construction, those with a portland cement base have the longest service record. There are two types of portland *cement paint:* Type I (containing a minimum of 65 percent portland cement by weight), for general use; and Type II (with at least 80 percent portland cement by weight), for maximum durability. Within each type there are two classes: Class A contains no aggregate filler and is for general use, whereas Class B contains 20 to 40 percent sand filler for use on open-textured surfaces.

A concrete masonry surface should be damp at the time of application of a portland cement paint. The setting and curing require the presence of water, a favorable temperature, and sufficient time for hydration. If the paint is modified with latex, moist curing is not necessary because the latex retains sufficient moisture in the paint film for hydration.

Although portland cement paints may be made on the job, the best results (uniform color and durability) are most often secured by using those marketed in prepared form. Portland cement paints form hard, flat porous films that readily permit passage of water vapor. These paints are not harmed by the presence of alkalies and may be applied to freshly erected concrete masonry surfaces. However, the results will be better if painting is deferred at least 3 weeks after new construction.

Latex Paints   Latex paints are water thinned, that is, they are based on aqueous emulsions of various resinous materials such as acrylic resin and polyvinyl acetate. With an ever increasing use of polymers, blends, and

Mortarless Wall Construction

A recent innovation in the masonry industry is the use of glass fibers as reinforcement in neat cement paste and mortar. Glass-reinforced cements and mortars are applied to dry-stacked masonry units with no mortar in the joints between units. Application is possible by hand or machine, and applying this surface-bonding material to both faces of the masonry unit promotes wall stability and strength. The surface-bonding material is applied in a $1/8$-inch thickness with plastering tools. Because of this relatively thin application, rodding or floating operations are unnecessary.

Commercial products are available with standard glass fibers, or alkali-resistant glass fibers. Some of the mixtures include a fine silica sand. The mixtures generally contain glass fibers that are approximately $1/2$ inch in length. Each fiber is a bundle of individual filaments held together and enveloped in a sizing material.

The structural capabilities of the glass-reinforced mixtures are being incorporated into pleasing architectural and structural components. Ask your local building supplier about this and other advances in masonry technology.

modifications of the base resins, latex bases are difficult to classify. But, they all dry very rapidly by evaporation of water, followed by coalescence of the resin particles.

Careful surface preparation is required for latex paints because they do not adhere readily to chalked, dirty, or glossy surfaces. However, these paints are easy to apply and have little odor. Also, they are economical, non-flammable, breathing paints that are not damaged by alkalies. They have excellent color retention and are very durable in normal environments. Those containing acrylics are more expensive than the other latex types, but experience has shown that they give the best performance.

Oil-Based and Oil-Alkyd Paints
Oil-based paints contain drying oils as

the binder and are nonbreathing. They are similar to conventional house paints. Easy to use, these paints are durable under some exposures but not particularly hard or resistant to abrasion, chemicals, or strong solvents. They are also damaged in the presence of alkalies.

Oil-based paints are often modified with alkyd resins to improve resistance to alkalies, reduce drying time, and improve performance in other ways. When the substitution of resins for oil is high, they are referred to as varnish-based paints. The oil-alkyds, and even the varnish-based paints, may be susceptible to damage from alkalies. Oil-alkyds also are nonbreathing paints. There are few instances where serious consideration should be given to an oil-based or oil-alkyd paint for use on concrete masonry.

Rubber-Based Paints These paints form a nonbreathing film and are alkali and acid resistant. They are used not only for exterior masonry surfaces but also for interior ones that are wet, humid, or subject to frequent washing (swimming pools, wash and shower rooms, kitchens, and laundries): that is, where alkali resistance is important and where requirements for resistance to the entrance of moisture are greater than can be supplied by latex paints. Rubber-based paints may be used as primers under less resistant paints.

Surface Preparation

Regardless of the type of paint selected, its success or failure can be dependent on the adequacy of surface preparation. In most cases success is assured if the concrete masonry surface can age at least 6 months before painting to get rid of the dampness and alkalinity which are characteristic of a new masonry wall. If the paint is not sensitive to either moisture or alkalies (such as a portland cement paint or latex paint), a long aging period is unnecessary. Earlier use of paint subject to damage from alkalinity is possible if the surface is neutralized by pretreatment with a 3-percent solution of phosphoric acid followed by a 2-percent solution of zinc chloride. However, this procedure has become rare because of the successful use of other paints that do not require pretreatment, such as portland cement and latex paints.

For paint to adhere, concrete masonry surfaces must be free of dirt, dust, grease, oil, and efflorescence. Dirt and dust may be removed by air-blowing, brushing, scrubbing, or hosing. If a surface is extremely dirty, it can be cleaned by means of wet or dry sand-blasting or power washing. Grease and oil are removed by applying a 10-percent solution of caustic soda, trisodium phosphate (TSP), or detergents specially formulated for use on concrete. Efflorescence is cleaned off by brushing, using a weak muriatic acid wash, or light sandblasting. After any of these treatments, the surface should be thoroughly flushed with clean water.

Fill coats, also called fillers or primer-sealers, are sometimes used to fill voids in open- or coarse-textured concrete masonry surfaces. Applied by brush before the finish coat(s), the fill coats usually contain white portland cement and fine sand. If acrylic latex or polyvinyl acetate latex is included in the mixture, moist curing is not required.

Paint Preparation and Application

Paint must be thoroughly stirred just prior to application. Power stirrers and automatic shakers are becoming more common, but they are not recommended for latex paints because of the possibility of foaming. Hand stirrers should have a broad, flat paddle.

Thinning of paint should only be done in accordance with the manufacturer's directions; excessive paint thinning will result in coatings of low durability. Color tinting should also be done carefully in accordance with the manufacturer's suggestions. The paints commonly used on concrete masonry are applied by brush, roller, or spray.

Portland cement paints are applied to damp surfaces by brush with bristles no more than 2 inches in length. The paint should be scrubbed into the surface. An interval of at least 12 hours should be allowed between coats. After completion of the final coat, a 48-hour moist curing is necessary if the paint is not modified with latex.

Latex paints may be applied to dry or damp surfaces by roller or spray, but preferably by a long-fiber, tapered nylon brush 4 to 6 inches wide (soaked in water for 2 hours before use). When the surface is moderately porous or extremely dry weather prevails, it is advisable to dampen the surface. These paints dry throughout as soon as the water of emulsion has evaporated, usually in 30 to 60 minutes, and require no moist curing.

Oil-based and oil-alkyd paints should not be applied during damp or humid weather or when the temperature is below 50°F. Application is by brush, roller, or spray (usually by brush) to a dry surface. Each coat should be allowed to dry at least 24 hours and preferably 48 hours before application of succeeding coats.

Rubber-based paints are usually applied by brush to dry surfaces. Two or three coats are necessary to achieve adequate film thickness, and the first

Brushes that are typically used in applying portland cement paint are (left to right) an ordinary scrub brush, a window brush, a brush with a detachable handle, and a fender brush.

The first coat of portland cement paint is brushed into the joints before it is applied to the rest of the block.

coat is usually thinned in accordance with the manufacturer's recommendations. A 48-hour delay is recommended between coats. Recoating should be performed with care. Because of the strong solvents used in rubber-based paints, a second coat tends to lift or attack the original coat.

CAUTION: Most paints are flammable, and some have solvents that are highly flammable. Therefore, adequate ventilation during painting should be provided according to the manufacturer's recommendations.

Clear Coatings

Clear coatings on concrete masonry surfaces render the surface water repellent and thus protect the masonry from soiling and surface attack by airborne pollutants, as well as facilitate cleaning. Further advantages are to prevent the surface from darkening when wet and to accentuate surface color. In some areas, where weather exposure is not severe and air pollution is low, coatings may not be necessary.

The coating selected should be water-clear and capable of being absorbed into the surface. It should also be long-lasting and not subject to discoloration with time and exposure. Check with a reputable paint dealer for a coating based on a methyl methacrylate form of acrylic resin. Brush or spray application of one or two coats of a relatively low solid content coating is usually satisfactory.

Stains

Decorative staining of concrete masonry walls can give good results if the proper stain is selected and then applied correctly. However, there may be some drawbacks to staining. For example, color applied after concrete hardens is not as long-lasting as that incorporated into the concrete mix dur-

ing block manufacture. Also, the color effects or shades of a stain may vary.

For best results, staining should be delayed at least 30 to 45 days after the concrete masonry structure is built. The surface must be dry and clean—free of oil, grease, paint, and wax. Two or more stain applications may be required to secure the depth of color desired.

Each coat of stain should thoroughly saturate the surface and be evenly applied by a constant number of passes of a brush, roller, or low-pressure spray. Care should be exercised so that the stain does not streak or overlap into a dried area. From 4 to 5 days should elapse between coats, depending on the masonry surface, conditions, and the stain used. It will often take 3 or 4 days for a stained surface to reach its final color.

Portland Cement Plaster Finishes

Portland cement *plaster* and portland cement *stucco* are essentially the same finish material. Although stucco is the term often associated with exterior use of the material, the term plaster does not explicitly denote interior use. Plaster is a combination of portland cement, masonry cement, or plastic cement with sand, water, and perhaps a plasticizing agent such as lime. A color pigment may be used in the finish coat.

Portland cement plaster has many of the desirable properties of concrete and, when properly applied, forms a durable, hard, strong, and decorative finish. There is an unlimited variety of textures, colors, and patterns possible. The rougher textures help conceal slight color variations, lap joints, uneven dirt accumulation, and streaking.

Portland cement plaster finishes are primarily used for exterior walls, but they are also particularly well suited for interior high-moisture locations such as kitchens, laundries, saunas, and bathrooms. The use of exterior plaster involves considerations of water penetration, corrosion of reinforcement and accessories, and stresses due to wider variations of temperature and humidity than normally present in interiors. Therefore, exterior applications require practices and precautions that are somewhat different than those for interior applications.

Plastering is probably one of the most difficult of all masonry skills, but knowing some of the basics about it may get you started. For a full discussion of portland cement plastering, refer to the *Portland Cement Plaster (Stucco) Manual*, available from the Portland Cement Association (see "For Further Reference," in the back of the book). A brief review of portland cement plaster follows.

Jointing for Crack Control

The proper design and selection of materials for a concrete masonry wall can substantially minimize or eliminate undesirable cracking of its portland cement plaster finish. Cracks can develop in this finish through many causes or combinations of causes: for example, shrinkage stresses; building movements; foundation settlements; restraints from lighting and plumbing fixtures, intersecting walls or ceilings, pilasters, and corners; weak sections due to cross-sectional changes at openings; and construction joints.

It is difficult to anticipate or prevent cracks from all these possible causes, but they can be largely controlled by means of control joints (see illustration 12-1). Whether plaster is applied directly to a concrete masonry base or on furred lath (metal reinforcement), control joints should be

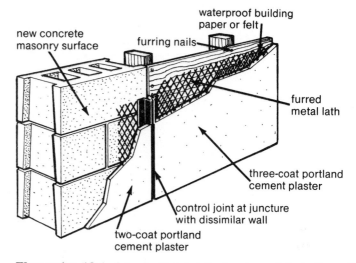

Illustration 12-1. A control joint at the juncture of dissimilar walls prevents random cracking. Three-coat plaster is needed on metal lath over wood construction.

installed directly over any previous joints in the base.

Concrete masonry walls that use metal lath for the plaster base should be divided into rectangular panels with a control joint at least every 20 feet. The metal reinforcement in the plaster must not extend across these control joints, and the material used for the control joints on exterior surfaces should be weathertight and corrosion resistant.

Mixes

A good plastering mix will be recognized by its workability, ease of troweling, adhesion to bases, and ability to attach itself to surfaces without sagging. Batch-to-batch uniformity will help ensure uniform suction for subsequent coats and color uniformity.

Data for plaster mixes are given in tables 12-1 and 12-2. Note that lime should not be added when masonry cement or plastic cement is used. These cements already contain plasticizers

How To Use These Tables

In Table 12-1 you will find the correct plaster mix combinations for the first two coats of plaster on two different types of wall base. The letter designations in Table 12-1 indicate plaster mix types that are recognized by the American National Standards Institute. You should use the mixes only in the combinations shown on the table. For example: If you are applying the plaster to metal reinforcement and if you choose to apply your scratch coat using Type CP plaster mix, you *must* use either Type CP or Type P as your brown coat.

Table 12-2 gives you the proportions of all ingredients necessary to make each type of plaster mix shown in Table 12-1. Note in Table 12-2 that the amount of sand called for in the different mixes depends on whether the plaster is for a scratch coat or a brown coat. To ensure uniformity of color, the finish coat should come from a factory-prepared stucco finish mix.

and only sand and water need be added, thus simplifying job-site proportioning and mixing.

Uniform measuring and batching methods are important. All ingredients should be thoroughly mixed (preferably with a power mixing attachment on a heavy-duty drill) with the amount of water needed to produce a plaster of workable consistency. Mixing time should be a minimum of 2 minutes after all materials are in the mixer, or until the mix is uniform in color. The size of a batch should be that which can be used immediately or in no more than $2\frac{1}{2}$ hours. Remixing, to restore plasticity with the addition of water, is permissible within the same time limits.

Color pigments are often used in the finish coat, which is usually a factory-prepared stucco finish mix. It should be noted that factory-prepared finish mixes assure greater uniformity of color than do-it-yourself mixes, and the manufacturer's recommendations should be closely followed.

Surface Preparation

Concrete masonry provides an excellent base for plaster because of its rigidity and excellent bonding characteristics. Bond occurs both mechanically and chemically. Mechanically, bond results from *keying;* that is, the interlocking of plaster with the open texture in the concrete masonry surface. Suction by the masonry also improves mechanical bond because plaster paste is drawn into minute pores of the surface. Chemically, the similar materials adhere well to each other.

A new concrete masonry surface can be used as a plaster base with minimum consideration. For best results, the concrete masonry units should have an open texture and be laid with struck joints. The surface should be free of oil, dirt, or other mate-

Table 12-1. **Permissible Mixes for Portland Cement Plaster Base Coats**

	Plaster Mix Symbols	
Type of Plaster Base	**First Coat (Scratch Coat)**	**Second Coat (Brown Coat)**
Concrete masonry*	L M P	L M P
Metal reinforcement†	C L CM M CP P	C, L, M, or CM L CM or M M CP or P P

*High-absorption bases such as concrete masonry should be moistened prior to scratch coat application.

†Metal reinforcement with paper backing may require dampening of paper prior to application of plaster.

Table 12-2. **Proportions for Portland Cement Plaster Base Coats**

Plaster Mix Symbols	Cementitious Materials, parts by volume				Sand*	
	Portland Cement	Lime*	Masonry Cement	Plastic Cement	First Coat	Second Coat†
C	1	0–3/4	—	—	not less than 2½ and not more than 4 times the sum of the volumes of cementitious materials used	not less than 3 and not more than 5 times the sum of the volumes of cementitious materials used
CM	1	—	1–2	—		
L	1	3/4–1½	—	—		
M	—	—	1	—		
CP	1	—	—	1		
P	—	—	—	1		

*Variations in lime and sand contents are given due to variations in local sands, and the fact that higher lime content will permit use of higher sand content. The workability of the plaster mix will govern the amounts of lime and sand.

†Within the limits shown, the same or greater proportions of sand should be used in the second coat as in the first coat.

rials that reduce bond. Then, just prior to application of plaster, it should be uniformly dampened, but not saturated, with clean water.

An old concrete masonry surface should be assessed to establish its bonding characteristics. A surface having the desired texture and cleanliness will perform as well as a new masonry surface. If the masonry has been painted, sandblasting may be the only way to remove the paint and improve the bonding characteristics. Otherwise, the base required must be obtained by anchoring metal lath to the surface over waterproof building paper or felt.

The suitability of a concrete masonry surface as a plaster base can be tested by spraying it with clean water to see how quickly moisture is absorbed through suction. If water is readily absorbed, good suction is likely; if water droplets form and run down the surface, its suction is probably inadequate. For low-suction surfaces, bond must be increased by applying a bonding agent or a dash-bond coat (containing 1 part

A fine water spray is used to dampen, but not saturate the concrete block surface in preparation for its first coat of plaster.

The scratch coat of plaster is applied by hand to metal lath that has been nailed over waterproof building paper or felt. Dimples in the metal lath provide proper furring for embedment of the metal.

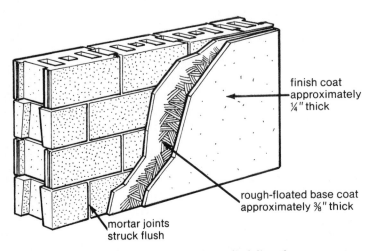

finish coat approximately ¼" thick

rough-floated base coat approximately ⅜" thick

mortar joints struck flush

Illustration 12-2. Two-coat plaster is applied directly to concrete masonry. Mortar joints on plaster-covered masonry walls are flush and the plaster is applied directly to the masonry surface.

Applying Plaster

Commercially, plaster and stucco are usually applied by machine, but you can apply plaster by hand. Whether plaster is applied by hand or machine, it must be applied with sufficient force to develop full adhesion between the plaster and the substrate and to put in place sufficient material to obtain the specified coat thickness. Plaster is applied in two or three coats in accordance with the required thicknesses given in table 12-3. Two coats are often used when plaster is applied directly to concrete masonry (see illustration 12-2). Three coats are applied when metal lath is used as a plaster base (see illustration 12-1).

The applied plaster must be brought to the desired thickness and the walls made plumb. The thickness of the coat of plaster is set by the placement of screed rails at the top and bottom of the area to be plastered. When the area between the screed rails is filled with mortar, a rod is used to even the surface. The rod can bear on the screed rails and be moved over the surface cutting off high spots and showing up the hollow places, which must be filled and rodded again. Additional manipulation of the surface is then required to prepare for the next coat.

Scratch-coat (first coat) plasters are scored or scratched to promote mechanical bond when the *brown coat* (second coat) is applied. Brown-coat plasters are applied and floated to even the surface and bring it to proper plane, provide a uniform suction throughout the *base-coat* plaster, and provide a desirable surface for the finish coat. Brown-coat surfaces are floated with a wooden float to improve bond with the final surface finish.

Hand Application The operation of applying plaster by hand begins when the plaster for the scratch coat is put on a mortarboard. First you ver-

portland cement, 1 to 2 parts sand, and sufficient water for a thick paintlike consistency). In lieu of a bonding agent or dash-bond coat, bond must be provided by metal lath over waterproof building paper or felt.

Table 12-3. **Thicknesses of Portland Cement Plaster**

Type of Work		Type of Plaster Base		Coat Thickness, in.							
				Coats on Vertical Surface				Coats on Horizontal Surface			
				1st	2nd	3rd*	Total	1st	2nd	3rd*	Total
Interior	three-coat work†	metal reinforcement		$1/4$	$1/4$	$1/8$	$5/8$	$1/4$	$1/4$	$1/8$	$5/8$
		solid base	unit masonry	$1/4$	$1/4$	$1/8$	$5/8$	use two-coat work			
			metal reinforcement over solid base	$1/2$	$1/4$	$1/8$	$7/8$	$1/2$	$1/4$	$1/8$	$7/8$
	two-coat work‡	solid base	unit masonry	$3/8$	$1/4$	—	$5/8$	—	—	—	$3/8$ max.§
Exterior	three-coat work†	metal reinforcement		$3/8$	$3/8$	$1/8$	$7/8$	$1/4$	$1/4$	$1/8$	$5/8$
		solid base	unit masonry	$1/4$	$1/4$	$1/8$	$5/8$	use two-coat work			
			metal reinforcement over solid base	$1/2$	$1/4$	$1/8$	$7/8$	$1/2$	$1/4$	$1/8$	$7/8$
	two-coat work‡	solid base	unit masonry	$3/8$	$1/4$	—	$5/8$	—	—	—	$3/8$ max.§

*The finish coat thickness may vary, provided that the total plaster thickness complies with this table and is sufficient to achieve the texture desired.

†Where three-coat work is required, dash or brush coats of plaster are not acceptable as one of the three coats.

‡For two-coat work, only the first and finish coats for vertical surfaces and the total plaster thickness for horizontal surfaces are indicated. The use of two coats is common practice when plaster is applied directly to vertical concrete masonry, and horizontal application seldom exceeds two coats.

§On horizontal solid-base surfaces such as ceilings or soffits requiring more than $3/8$ inch plaster thickness to obtain a level plane, metal reinforcement should be attached to the concrete masonry and the thickness specified for three-coat work on metal reinforcement over solid base applies. Where $3/8$ inch or less plaster thickness is required to level and decorate and there are no other requirements, a liquid-bonding agent or dash-bond coat may be used.

ify that the plaster is properly mixed by kneading the plaster on the board. Then take some plaster from the board, put it on the hawk and begin to plaster the wall surface. After lifting the plaster from the hawk onto the trowel, lay the plaster on the wall surface. Plastering can be done from the bottom to the top of the work area, or from top to bottom, according to your preference.

Use enough pressure to obtain good contact between plaster and base surface. This procedure continues until the entire wall is plastered to the desired thickness. Excessive troweling or movement of the scratch coat must be avoided as it is being applied, because too much action will break

The brown coat is applied over the scratch coat and made thick enough to bring the surface to the proper plane.

ter has been applied, the surface is rodded to the desired plane. The plaster thickness is properly gauged with plaster screeds or wooden slats of proper thickness as the guides. These wooden slats are tacked on either side of the stuccoed area and removed after the stuccoing operation is completed.

After rodding, float the surface to give it the correct surface texture. Floating of the brown coat is the most important part of plastering. Floating must be done only after the plaster has lost sufficient moisture so that the surface sheen has disappeared and before the plaster has become so rigid that it cannot be moved under the float. This interval is critical, because the degree of consolidation that occurs during floating influences the shrinkage-cracking characteristics of the plaster.

A few minutes after the brown coat is applied, the surface is rodded to the desired plane.

the bond created between the plaster and the base, whether masonry or metal lath.

The scratch coat should be scored in a horizontal direction. Shallow scratching is adequate.

The brown coat is applied next in sufficient thickness, usually $3/8$ to $1/2$ inch, to bring the surface to the proper plane. A few minutes after the plas-

The finish coat is applied to a predampened, but still absorptive, brown coat to a thickness of about $1/8$ inch. The finish coat is applied from the top down and the whole wall surface must be covered without laps or interruptions.

Many types of textures may be obtained with a trowel. A highly textured surface may also be obtained using a dash brush. A dash brush has bristles about 4 inches long and is used by loading it with mix and flipping the mix onto the brown coat.

Curing Portland cement plaster requires moist curing after application in order to produce a strong, durable finish. Curing can be successfully accomplished by several methods; the type and size of the structure and the climatic and job conditions will dictate the one that is most suitable. For example, plaster can be moist cured by using a suitable covering, such as building plastic, to provide a vapor barrier that will retain the moisture in the plaster. Moist curing can also be accomplished by using a fine spray of water. Tarpaulins or plywood barriers that deflect sunlight and wind will help to reduce evaporation rates.

An adequate temperature level is important to satisfactory curing. As the temperature drops, hydration slows and practically stops when the temperature approaches the freezing point. Therefore, portland cement plaster should not be applied to frozen surfaces, and frozen materials should not be used in the mix. Furthermore, in cold weather it may be necessary to heat the mixing water and the work area.

Furred Finishes

In addition to paint or plaster, other finishes such as fiberboard, gypsum wallboard, and wood paneling are sometimes applied directly to concrete masonry wall surfaces and sometimes

The finish coat is applied to a predamped, but still absorptive, brown coat and floated and textured to the desired finish.

on furring. The furring, which may be wood or metal, ensures a definite air space, from $3/4$ inch to several inches wide, between the masonry and the finish. Furring may be necessary to accomplish the following tasks.

• Provide suitably plumb, true, and properly spaced supporting construction for a wall finish.

• Eliminate capillary moisture transfer in exterior or below-grade concrete masonry walls, thus minimizing the likelihood of condensation on interior wall surfaces.

• Improve thermal insulation.

• Improve sound insulation.

The furring strips are fastened to the concrete masonry with cut nails, helically threaded concrete nails, or powder-actuated fasteners. Adhesives can be used to attach wallboard directly to wooden furring, although a few nails may be required until the adhesive has set.

Gypsum wallboard finishes (drywall) consist of one or more plies of factory-fabricated gypsum board hav-

Wooden furring strips are nailed to the mortar joints between the blocks.

ing noncombustible gypsum cores with surfaces and edges of paper. Gypsum plaster may be applied directly to concrete masonry or to lath that may or may not be furred. Gypsum products are not recommended where significant exposure to moisture is expected.

Typical materials used to increase thermal resistance of furred finishes include the following:

• Fiberglass batts or blankets

• Treated cellulose or other loose-fill insulation

• Rigid polystyrene panels

• Reflective foil

Damp Proofing

Condensation in basements is caused by precipitation of moisture from warm, moist air coming in contact with the relatively cool surfaces of basement walls or floors. Inadequate drainage is one of the primary causes of leakage into basements, and if there is standing water around a house, leakage is more apt to occur. Lack of drainage in the ground, from the yard, or away from the house may be to blame. The other major cause of leakage is inferior construction. Here are some prime examples:

• Imperfectly laid masonry with inadequately filled and compacted mortar joints or poor-quality mortar

• Premature backfilling against unbraced walls, causing cracking

• Failure to waterproof the outside of concrete masonry basement walls

• Inadequate footings, which may result in wall cracks due to settling

• Lack of control joints

Using good masonry practices as discussed in Chapter 11 is the first step in creating damp-proof walls. To provide further protection, coatings are required. The earth side of concrete masonry basement walls should be covered with a ½-inch-thick coat of plaster, preferably applied in two layers. Either portland cement plaster (1 part cement to 2½ parts sand by volume) or the mortar used for laying up the block can be used. However, portland-cement-based coatings that have been specifically prepared to waterproof masonry basement walls are preferred over the plain plaster coat.

The wall surface should be cleaned and dampened (not soaked) with water, preferably by spraying before the application of the plaster. This will prevent the block from absorbing excessive water from the plaster and will ensure better bond.

When the plaster is to be applied in two coats, the first coat should be parged (troweled) firmly over the masonry. When the plaster has hardened partially, the surface should be roughened with a scratcher just as you would for regular plastering. The first coat should be kept damp and allowed to harden for at least 24 hours before applying the second coat. The plaster should extend from the top of the wall down to the footing, where the plaster is coved to prevent water from collecting around the juncture of wall and footing. Just before application of the second coat, the roughened surface should be dampened with water (not soaked) for good bond. The second coat should be moist cured for at least 48 hours after application.

In poorly drained, wet soils, the plaster coating should be covered with two thick, cement-sand grout coats scrubbed well into the surface with an ordinary floor scrub brush or other stiff-bristled brush. This treatment can also be applied to porous, cast-in-place concrete walls. The plaster of the wall should be dampened before applying each coat of grout. Each coat should be cured by being kept wet for at least 24 hours.

Various methods can be used to damp proof concrete masonry basement walls, depending on climate and subsurface drainage.

Two coats of cement-based paint, prepared specifically for waterproofing foundation walls, or a waterproof membrane can be used instead of the grout.

CHAPTER 13

Masonry Projects

The most common application of concrete masonry is for built-in-place walls used on buildings of all kinds. However, there are a number of other common applications: such as fireplaces, retaining walls, walks, and patios. Whatever the application, successful masonry projects require planning. This planning should include making detailed drawings or sketches, assessing material needs, and obtaining the required building permits. If you live in a temperate climate, the ideal time for you to plan outdoor projects is during the winter and spring so construction can take place during the spring and summer when conditions are ideal.

Fireplaces and Chimneys

Fireplaces and chimneys are important elements in the design and construction of many homes. The fireplace can be a central feature for family social life, and the chimney often is a dominant and interesting architectural feature on the exterior of the home. In

short, fireplaces and chimneys must be both functional and aesthetically pleasing.

Requirements for fireplaces and chimneys are normally set forth in local building codes. Usually, they pertain to a single residential fireplace with the chimney tied to the roof or ceiling rafters. In the event that the chimney is multistory, extrawide, or extrahigh —or there are multiple fireplaces and flues within the chimney—special design considerations are necessary.

The design and construction of an efficient, functional fireplace requires adherence to basic rules concerning fireplace location and the dimensions and placement of various component parts. A good design will incorporate elements both of simplicity and safety, and will enhance the basic functions of a fireplace, which are as follows:

- To assure proper fuel combustion
- To deliver smoke and other products of combustion up the chimney

• To radiate the maximum amount of heat

• To provide an attractive architectural feature

Combustion and smoke delivery depend mainly on the shape and dimensions of the combustion chamber, the proper location of the fireplace throat and the smoke shelf, and the ratio of the flue area to the area of the fireplace opening. The third objective, heat radiation, depends on the dimensions of the combustion chamber. Fire safety depends not only on the design of the fireplace and chimney, but also on the ability of the masonry units to withstand high temperatures without warping, cracking, or deteriorating.

Barbecues or outdoor fireplaces can discharge into a chimney attached to the house with a separate flue, or if desired, can be located separately from the house with its own chimney. Inexpensive, serviceable barbecues can be built of masonry with minimum labor and time (see illustration 13-1). But there is no limit to design of barbecues and you may want a more elaborate structure (see illustration 13-2). The site selected for any barbecue should be sheltered from the wind, conveniently located between play and work areas, and afford adequate entertaining space.

Fireplace Elements

The details of a typical unreinforced concrete masonry fireplace are shown in illustration 13-3. We will now describe the major elements that make up a fireplace.

The floor of the fireplace is called the *hearth*. The inner part of the hearth is lined with firebrick and the outer hearth consists of noncombustible material such as firebrick, brick, concrete block, or concrete. The outer

Fireplaces are important from both a design and function standpoint, and care must be taken in considering both when constructing one.

hearth is supported on concrete that may be part of a ground floor or a cantilevered section of the slab supporting the inner hearth.

The fireplace *lintel* is the horizontal member that supports the front face or breast of the fireplace above the opening. It may be made of reinforced masonry or steel angle.

The combustion chamber where the fire occurs is called the *firebox*. Its sidewalls are slanted slightly to radiate heat into the room, and its rear wall is curved or inclined to provide an upward draft to the throat.

Unless the firebox is of the preformed metal type (at least $1/4$ inch thick), it should be lined with firebrick at least 2 inches thick, laid with joints of fireclay mortar. Its back and sidewalls, including lining, should be at least 8 inches thick to support the weight of the chimney above.

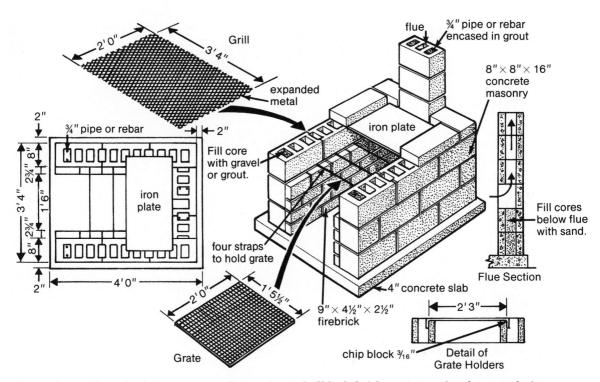

Illustration 13-1. A simple barbecue may be constructed of block, brick, or stone using the same design techniques shown here.

In the generally accepted method of construction on a concrete slab, the fireplace is laid out and its back constructed to a scaffold height of approximately 5 feet before the firebox is constructed and backfilled with tempered mortar and brick scraps. You should not backfill solidly behind the firebox wall but slush the mortar loosely to allow for some expansion of the firebox.

The *throat* of a fireplace is the slotlike opening directly above the firebox through which flames and smoke pass into the smoke chamber. Because of its effect on the draft, the throat must be carefully designed to be not less than 6 inches, and preferably 8 inches, above the highest point of the fireplace opening (see illustration 13-3).

The sloping or inclined back of the firebox should extend to the same height as the throat and form the support for the hinge of a metal *damper* placed in the throat. The damper extends the full width of the fireplace opening and preferably opens both upward and backward.

The *smoke chamber* acts as a funnel to compress the smoke and gases from the fire so that they will enter the chimney flue above. The shape of this chamber should be symmetrical with respect to the centerline of the

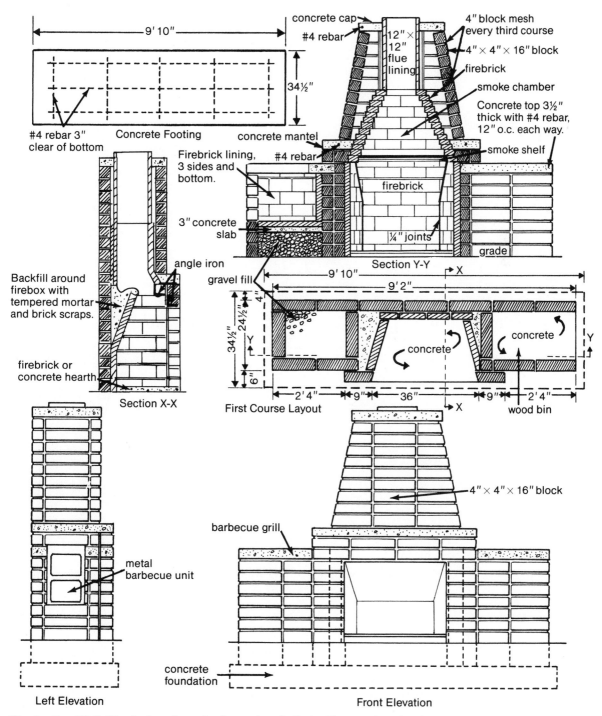

9' 10"

#4 rebar 3"
clear of bottom

Concrete Footing

concrete cap
#4 rebar

12" × 12" flue lining

4" block mesh
every third course

4" × 4" × 16" block

firebrick

smoke chamber

Concrete top 3½"
thick with #4 rebar,
12" o.c. each way.

34½"

concrete mantel

#4 rebar

smoke shelf

Firebrick lining,
3 sides and
bottom.

firebrick

3" concrete
slab

¼" joints

grade

Section Y-Y

angle iron

gravel fill

Backfill around
firebox with
tempered mortar
and brick scraps.

9' 10"

9' 2"

concrete

concrete

firebrick or
concrete hearth.

34½" 24½" 4"

6"

2' 4" 9" 36" 9" 2' 4"

wood bin

Section X-X

First Course Layout

metal
barbecue unit

barbecue grill

4" × 4" × 16" block

concrete
foundation

Left Elevation

Front Elevation

Illustration 13-2. The design of your barbecue is only limited by your imagination.

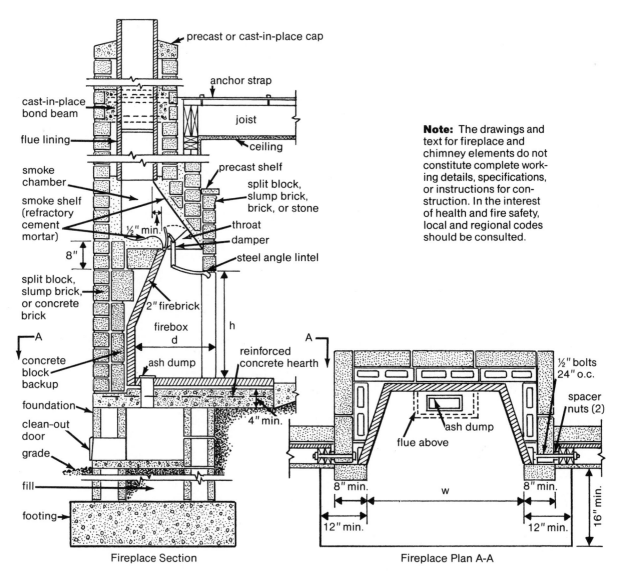

Note: The drawings and text for fireplace and chimney elements do not constitute complete working details, specifications, or instructions for construction. In the interest of health and fire safety, local and regional codes should be consulted.

Fireplace Section   Fireplace Plan A-A

Illustration 13-3. This is an example of an unreinforced masonry fireplace and chimney. Some building codes require that concrete masonry units used in chimney and fireplace construction be solid.

firebox in order to ensure even burning across the width of the firebox. The back of the smoke chamber is usually vertical and its other walls are inclined upward to meet the bottom of the chimney flue lining. If the wall thickness is less than 8 inches of solid masonry, the smoke chamber should be parged with a 3/4-inch-thick coat of fireclay mortar. Metal lining plates are available to give the chamber its proper form, provide smooth surfaces, and simplify masonry construction.

Chimney Elements

A fireplace chimney serves a dual purpose: it creates a draft and disposes of the products of combustion. Careful consideration must be given to chimney design and erection in order to assure efficient operation and freedom from fire hazards.

To prevent upward draft from being neutralized by downward air currents, the chimney should be built at least 3 feet above a flat roof, 2 feet above the ridge of a pitched roof, or 2 feet above any part of the roof within a 10-foot radius of the chimney. If the chimney does not draw well, increasing its height will improve the draft.

Usually made of concrete, the *foundation* for a chimney is designed to support the weight of the chimney and any additional load, such as the fireplace and floors. Because of the large mass and weight imposed, it is important that the unit bearing pressure beneath the chimney foundation be approximately equal to that beneath the house foundation; this will minimize the possibility of differential settlement. The chimney foundation is generally unreinforced, with only the chimney reinforcement (where required by local building codes) extending from it.

The *footing* thickness should be at least 8 inches and not less than $1\frac{1}{2}$ times the footing projection, unless reinforced. The bottom of the footing should be at least 18 inches below grade and extend below the frost line.

A fireplace chimney *flue* must have the correct area and shape to produce a proper draft. Relatively high velocities of smoke through the throat and flue are desirable. Velocity is affected by the flue area, the firebox opening area, and the chimney height.

Generally the required cross-sectional area of the flue should be approximately $\frac{1}{10}$ of the area of the fireplace opening. However, because some codes may specify $\frac{1}{8}$ or $\frac{1}{12}$ under varying conditions, the local building department should be consulted. Typical sizes of fireplace openings and flue linings are given in tables 13-1 and 13-2.

A fireplace chimney can contain more than one flue, but each flue must be built as a separate unit entirely free from the other flues or openings. Flue walls should have all joints completely filled with mortar. All chimney flues should be lined, and clay flue liners are the common requirement.

Flue linings should start at the top of the smoke chamber and extend continuously until 4 to 8 inches above the chimney cap. The chimney walls are constructed around the flue lining segments, which are embedded one on the other in a refractory mortar, such as fireclay, and left smooth on the inside of the lining. Liners should

Table 13-1.	**Standard Sizes for Single-Face Fireplaces**			
Width (w), in.*	Height (h), in.*	Depth (d), in.*	Area of Fireplace Opening, sq. in.	Nominal Flue Sizes (based on $\frac{1}{10}$ area of fireplace opening), in.†
36	26	20	936	12×16
40	28	22	1,120	12×16
48	32	25	1,536	16×16
60	32	25	1,920	16×20

*For examples of how to measure w, h, and d, see illustrations 13-3 and 13-5.

†A requirement of the U. S. Federal Housing Administration if the chimney is 15 feet high or over; $\frac{1}{8}$ ratio is used if chimney height is less than 15 feet.

Table 13-2. **Clay Flue Lining Sizes**

Nominal Size, in.	Manufactured Size (modular), in.*	Inside Area, sq. in.
4 × 8	3$\frac{1}{2}$ × 7$\frac{1}{2}$	15
4 × 12	3$\frac{1}{2}$ × 11$\frac{1}{2}$	20
4 × 16	3$\frac{1}{2}$ × 15$\frac{1}{2}$	27
8 × 8	7$\frac{1}{2}$ × 7$\frac{1}{2}$	35
8 × 12	7$\frac{1}{2}$ × 11$\frac{1}{2}$	57
8 × 16	7$\frac{1}{2}$ × 15$\frac{1}{2}$	74
12 × 12	11$\frac{1}{2}$ × 11$\frac{1}{2}$	87
12 × 16	11$\frac{1}{2}$ × 15$\frac{1}{2}$	120
16 × 16	15$\frac{1}{2}$ × 15$\frac{1}{2}$	162
16 × 20	15$\frac{1}{2}$ × 19$\frac{1}{2}$	208
20 × 20	19$\frac{1}{2}$ × 19$\frac{1}{2}$	262

*Actual dimensions may vary somewhat, but the flue lining must fit into a rectangle corresponding to the nominal flue size.

be separated from the chimney wall and the space between the liner and masonry is not filled; only enough mortar should be used to make a good joint and hold the liners in position.

When a chimney contains more than two flues, they should be separated into groups of one or two flues by 4-inch-thick masonry bonded into the chimney wall, or the joints of the adjacent flue linings should be staggered at least 7 inches. The tops of the flues should have a height difference of 2 to 12 inches to prevent smoke from pouring from one flue into another. A fireplace on an upper level should have the top of its flue higher than the flue of a fireplace on a lower level.

Chimneys should be built as nearly vertical as possible, but a slope is allowed if the full area of the flue is maintained throughout its length. When a slope from the vertical is required in the flue for design reasons, it should not exceed 7 inches to the floor or 30 degrees. Where offsets or bends are necessary, they should be formed by mitering both ends of abutting flue liner sections equally. This prevents reduction of the flue area.

The top of the chimney wall should be protected by a concrete *cap* conforming with the architectural design of the building. The cap should slope not only to prevent water from running down next to the flue lining and into the fireplace but also to prevent standing water from creating frost or moisture problems. In addition, because chimney flues should project at least 2 inches above the cap, a sloping cap improves draft from the flue as well as the smoke exhaust characteristics of the chimney. If the cap projects beyond the chimney wall a few inches, a drip slot in its lower edge should be included to help keep the wall dry and clean.

A chimney *hood* gives a finished touch to the silhouette of the building. It protects the chimney and fireplace from rain and snow and—when the building is located below adjoining

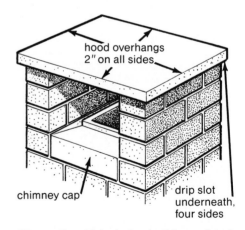

Illustration 13-4. A simple chimney hood keeps rain and snow out, prevents downdrafts, and improves the appearance of a chimney.

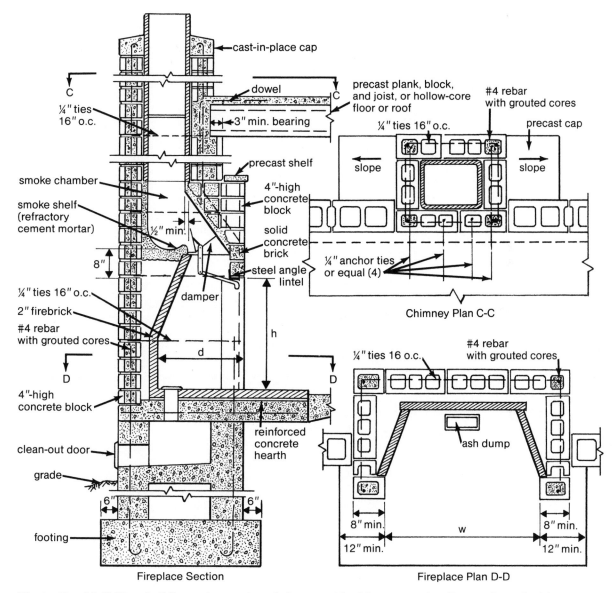

Illustration 13-5. Many building codes require reinforcement in chimney construction, as shown in this example.

buildings, trees, and other obstacles—prevents downdrafts. It must have at least two sides open, with the open areas larger than the flue area (see illustration 13-4).

Concrete chimney hoods should be reinforced with steel bars or welded-wire fabric. If a hood projects from a chimney wall, a drip slot under the edge is included. Also, the openings are sometimes enclosed with heavy screening to keep out small animals and birds. But, check your insurance company's regulations on screening.

Reinforcement and Chimney Anchorage

Depending on local building codes, fireplaces and chimneys have to be reinforced (see illustration 13-5) in areas subject to earthquakes or high wind loads.

The reinforcement, consisting of at least four 1/2-inch-diameter deformed vertical bars, should extend the full height of the chimney and be tied into the footing and chimney cap. Also, the bars should be tied horizontally with 1/4-inch-diameter ties at not more than 18-inch intervals. If the width of a reinforced chimney exceeds 40 inches, two additional 1/2-inch vertical bars should be provided for each additional flue incorporated into the chimney, or for each additional 40-inch width or fraction thereof.

All chimneys not located entirely within the exterior walls of a residence should be anchored to the building at each floor or ceiling line 6 feet or more above grade and at the roofline. The anchors should consist of 1/4-inch steel straps wrapped around vertical reinforcement or chimney flues (see illustration 13-3). Each end of the strap is attached to the structural framework of the building with six 16d nails, two 1/2-inch-diameter bolts, or two 3/8 × 3-inch lag screws. Reinforced chimneys must have equivalent anchorage (see illustration 13-5). When a chimney extends considerably above the roof level, an intermediate lateral support or tie is often placed between the roofline and the chimney.

Screen Walls

Concrete masonry screen walls are functional, decorative building ele-ments. They combine privacy with a view, interior light with shade and solar heat reduction, and airy comfort with wind control. Curtain walls, sun screens, decorative veneers, room dividers, and fences are just a few of the many applications of the concrete masonry screen wall.

Materials

With conventional concrete block or the specially designed screen wall units or grille block, concrete masonry offers a new dimension in screen wall design. The many sizes, shapes, colors, patterns, and textures available make imaginative designs easy to create. Several units may make up a design, or each screen wall unit may constitute a design in itself.

Although the number of designs for concrete masonry screen units is virtually unlimited, it is advisable to check on availability of any specific unit during the early planning stage. Some designs are available only in certain localities and others are restricted by patent or copyright.

Screen units should be of high quality. Type N mortar should be used for exterior screen walls aboveground and where the screen walls are required to carry any vertical load. For interior non-load-bearing walls, the mortar may be Type S or N. Reinforcement and grout should be used as required by local code.

Design and Construction

Although from a design standpoint, screen walls are seldom required to support more load than their own weight, care must be taken to see that they are stable and safe. If you want to design a screen wall to be load bearing, check your local building codes first, because some codes forbid this. In any case, extra attention to design of screen

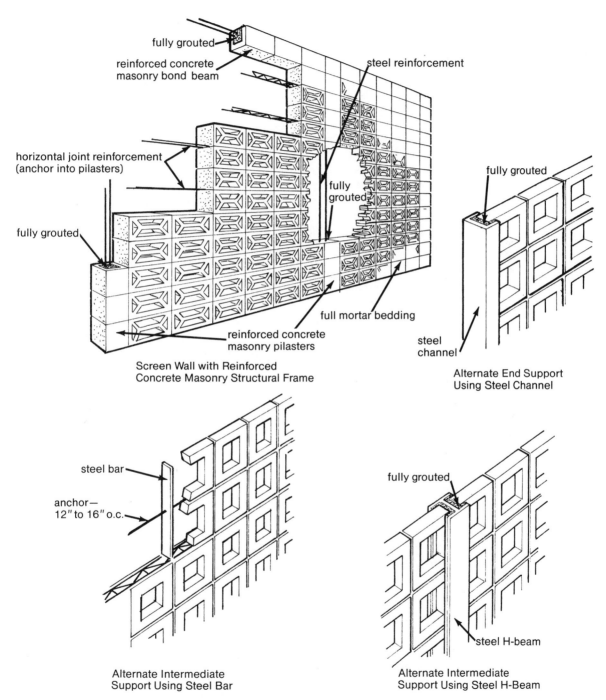

fully grouted

reinforced concrete
masonry bond beam

steel reinforcement

horizontal joint reinforcement
(anchor into pilasters)

fully
grouted

fully grouted

full mortar bedding

reinforced concrete
masonry pilasters

Screen Wall with Reinforced
Concrete Masonry Structural Frame

fully grouted

steel
channel

Alternate End Support
Using Steel Channel

steel bar

anchor—
12" to 16" o.c.

Alternate Intermediate
Support Using Steel Bar

fully grouted

steel H-beam

Alternate Intermediate
Support Using Steel H-Beam

Illustration 13-6. There are several different ways to reinforce screen walls because a single type of reinforcement will not work with every style of wall design.

walls for wind forces is warranted despite the relatively high percentage of open areas.

Screen walls should be designed to resist all the horizontal forces that can be expected. Stability is provided in the following ways.

• Use a framing system capable of carrying horizontal forces into the ground.

• Employ adequate connection or anchorage of screen walls to the framing system.

• Limit the clear span.

• Incorporate vertical reinforcement and horizontal joint reinforcement.

Partitions built with screen units are usually designed as non-load-bearing panels, with primary consideration given to adequate anchorage at panel ends and/or top edge, depending on where lateral support is furnished. Lateral support for screen walls may be obtained from cross walls, piers, columns, posts, or buttresses for the horizontal spans—and from floors, shelf angles, roofs, bond beams, or spandrel beams for screen walls spanning the vertical direction. The structural frame for a screen wall may consist of reinforced concrete masonry columns, pilasters, and beams, or may incorporate structural steel members (see illustration 13-6). Screen wall framing methods may also be similar to those used for fences (described in the next section).

When designed as veneer, the concrete masonry screen wall is attached to a structural backing with wire ties or sheet metal anchors in the same manner as used for other types of masonry veneer.

A non-load-bearing screen block panel may be used to fill an opening in a load-bearing masonry wall. In this case the panel is restrained on all four sides. Joint reinforcement is placed in the horizontal joints to anchor the panel into the wall. For an exterior wall, a panel is limited to 144 square feet of wall surface or 15 feet in any direction. The lintel, sill, and jamb of the panel opening should be designed the same as for a window opening.

Non-load-bearing screen walls should have a minimum nominal thickness of 4 inches and a maximum clear span of 36 times the nominal thickness. For load-bearing screen walls, the minimum thickness should be increased to 6 inches. The maximum span can be measured vertically or horizontally, but need not be limited in both directions. Screen walls are usually capable of carrying their own weight up to 20 feet in height, but above that height they must be supported vertically not more than every 12 feet.

Due to the somewhat fragile nature of screen walls, the use of steel reinforcement is recommended wherever it can be embedded in mortar joints and bond-beam courses, or grouted into continuous vertical or horizontal cavities. When reinforced joints are used, the thickness of the mortar joint should be a minimum of twice the diameter of the reinforcement.

For exterior screen walls, joints and connections should be constructed as watertight as possible. The mortar joints should be made according to the best construction practices. In addition, when hollow masonry units are laid with their cores vertical, the top course should be capped to prevent the entrance of water into the wall interior.

Garden Walls and Fences

Garden walls and fences of concrete masonry can take on many delightful forms, enhancing the landscape. They are built with solid or screen block and with brick or half block. If a garden wall has more than 25 percent open areas, it may be considered a fence. Fences are framed using a variety of methods.

Fences and garden walls should be able to safely withstand wind loads of at least 5 pounds per square foot (psf), and most city codes specify resistance to 20-psf pressure. Pressures and corresponding wind gust velocities are shown in table 13-3. In hurricane areas higher wind-pressure resistance is needed.

Reinforcement for walls is based on the pressure resistance required, distance between pilasters, and the size of reinforcing steel. Without reinforcement, high and straight garden walls or fences lack vertical tensile strength and are unstable in strong winds. For a 57-mph peak wind velocity the safe height of a straight 6-inch-thick block wall is only 3^1/$_2$ feet; for a straight 8-inch-thick wall it's only 5^1/$_2$ feet.

The serpentine wall or fence is a welcome change from the straight lines frequently seen on our modern landscapes. Undulating curves give this type of wall a stability from the foundation up with no need for reinforcement.

Illustration 13-7 shows sample designs of serpentine walls based on proportions found safe for wind gusts with pressures up to 20 psf; the horizontal radius should not exceed twice the height, which in turn should not be more than twice the width. A limiting height of 15 times the thickness is recommended. The free end(s) of the serpentine wall should have additional support, such as a pilaster or a short-radius return.

Concrete masonry wall foundations may not be durable if they frequently become frozen while saturated. In cold climates, therefore, the wall foundations should be constructed with cast-in-place concrete.

Brick Garden Walls

Brick garden walls are built using the same construction practices used in concrete masonry construction with a few exceptions depending on the type of wall you want to build. Two common types of brick wall are veneered and solid brick. Solid brick walls may be either straight, pier and panel, or serpentine.

Building a veneered brick wall requires building a concrete masonry unit wall in the manner described, then laying a veneer of brick to one or both sides. The veneer is secured to the wall using metal ties or reinforcement.

Straight walls require a footing poured level with or just below ground level. Straight walls consist of two wythes that are bonded either with reinforcement or by using bonding courses. Begin construction by striking chalk lines on the footing and laying out the first course without mortar to check spacing. Construct the wall using standard masonry techniques.

A wall requires a *coping* (cap) to prevent moisture penetration. The coping may consist of special brick units, stone, or precast concrete slabs. This

Table 13-3. **Wind Gust Pressures**	
Pressure, lbs. per sq. ft.	**Wind Gust Velocity, m.p.h.**
5	40
10	57
15	69
20	80

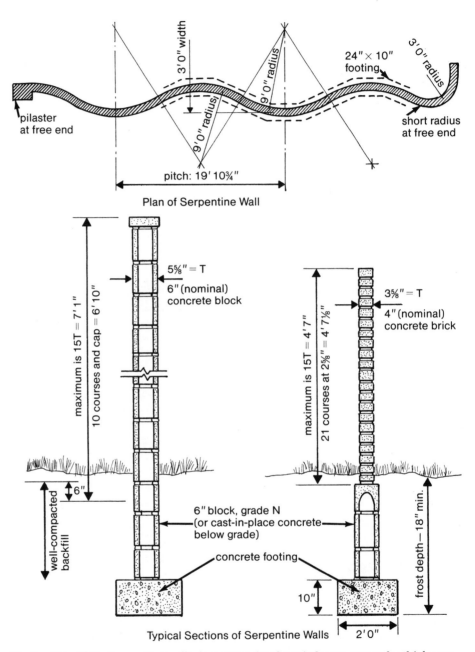

Illustration 13-7. Serpentine walls do not require the reinforcement or the thickness of a straight wall of comparable height.

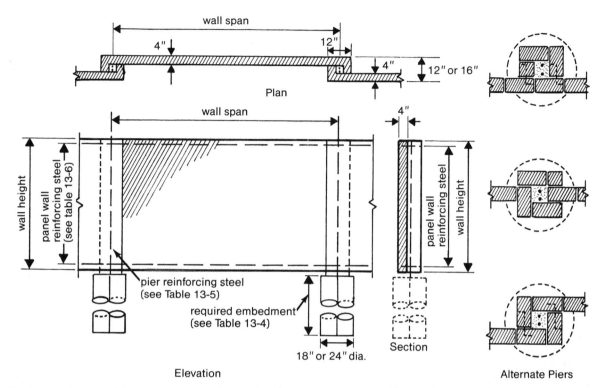

Illustration 13-8. Brick walls using panel-and-pier construction are built differently than those with a continuous foundation, but their design is still based on wind load, wall height, and pier depth.

coping must project beyond the edge of the wall at least ½ inch on each side in order to keep water from flowing down the face of the wall.

Another type of brick wall construction that is commonly used is the pier-and-panel wall (see illustration 13-8). In this type of construction, the walls are only one wythe thick, but they are built between reinforced piers. The piers, spaced 8 to 16 feet apart, are built on 18-inch-diameter footings that are at least 2 to 3 feet deep depending on wind load (see table 13-4). Continuous reinforcement extends from the base of the piers to the top of the wall and their size is noted in table 13-5. In addition to the verti-

			Table 13-4. **Required Embedment for Pier Foundation**			

	Embedment					
	Wind Load, 10 lbs. per sq. ft.		Wind Load, 15 lbs. per sq. ft.		Wind Load, 20 lbs. per sq ft.	
	Wall Height, ft.		Wall Height, ft.		Wall Height, ft.	
Wall Span, ft.	**4**	**6**	**4**	**6**	**4**	**6**
8	2'0"	2'3"	2'3"	2'6"	2'3"	2'9"
10	2'0"	2'6"	2'3"	2'9"	2'6"	3'0"
12	2'3"	2'6"	2'3"	3'0"	2'6"	3'3"
14	2'3"	2'9"	2'6"	3'0"	2'9"	3'3"
16	2'3"	2'9"	2'6"	3'3"	2'9"	3'3"

Table 13-5. **Pier Reinforcing Steel***

Wall Span, ft.	Wind Load, 10 lbs. per sq. ft. Wall Height, ft.		Wind Load, 15 lbs. per sq. ft. Wall Height, ft.		Wind Load, 20 lbs. per sq ft. Wall Height, ft.	
	4	6	4	6	4	6
8	#3	#4	#3	#5	#4	#5
10	#3	#4	#4	#5	#4	#6
12	#3	#5	#4	#6	#4	#6
14	#3	#5	#4	#6	#5	#5
16	#4	#5	#4	#6	#5	#6

*In each case, the designated rebar should be used in pairs.

Table 13-6. **Horizontal Reinforcing Steel for Panel Walls**

Wall Span, ft.	Wind Load, 10 lbs. per sq. ft.			Wind Load, 15 lbs. per sq. ft.			Wind Load, 20 lbs. per sq. ft.		
	A	B	C	A	B	C	A	B	C
8	45	30	19	30	20	12	23	15	9.5
10	29	19	12	19	13	8.0	14	10	6.0
12	20	13	8.5	13	9.0	5.5	10	7.0	4.0
14	15	10	6.5	10	6.5	4.0	7.5	5.0	3.0
16	11	7.5	5.0	7.5	5.0	3.0	6.0	4.0	2.5

NOTE: A = two #2 bars;
B = two 3/16 inch diameter wires
C = two 9 gauge wires

ting the reinforcing rods. Temporary wooden forms are set on the ground between the piers, and the walls and piers are constructed at the same time. Reinforcement is embedded as work progresses and grout made of the same mortar used on the walls is cast in the piers to secure the reinforcement.

Serpentine walls are constructed with brick in the identical method used for concrete masonry units (see illustration 13-7).

Stone Walls

Stone walls are constructed either with or without mortar. Dry construction uses no mortar and the stones are held together by friction and gravity (see illustration 13-9). The first step in this construction is to dig down 6 inches below grade and lay the first course. Use your largest stones for the first course not only to create a good base but also to avoid lifting and adjusting heavy stones at higher levels. Walls over 2 feet high should be *battered* (slanted) at a rate of 1/2 inch per foot of rise.

As you construct a dry stone wall, it is important to check the fit of each stone as you place it. If necessary, use smaller stones as chink in order to create a good base for the next stone. Avoid straight lines, both vertical and horizontal, by overlapping and staggering the stones as much as possible. Check the slope as you build and save the flattest stones in your pile for the cap. If you live in an area of severe freezing, consider mortaring the cap in place.

Wet stone construction relies on mortar to keep the stones in place and the wall is built on a footing to protect it from ground heave caused by freezing and thawing. Begin construction of a wet wall by making a footing that is as wide as the base of the wall and

cal reinforcement in the piers, horizontal reinforcement in the walls is required as indicated (see table 13-6).

Construction of a pier-and-panel wall begins by digging and pouring the piers below the frost line and set-

extending 6 inches below the frost line. The mortar for a wet stone wall is a little different than mortar used for brick or masonry units in that it contains no lime. The mortar should consist of 1 part portland cement to 3 or 4 parts sand, and you can add ½ part fireclay for workability. In addition, stones should be clean and dry before laying.

Begin construction by dry arranging the first course on the footing. As with dry wall construction, reserve the largest stones for the first course and the flattest ones for the cap. Stones are heavier than brick or masonry units. Because of this you need to use a stiff mortar mix, small stones and chips between stones, and wedges to support the stones while the mortar sets up (see illustration 13-10). As you work, clean up any spilled mortar with a wet sponge.

Retaining Walls

Concrete masonry retaining walls can have visual beauty along with the required structural strength to resist imposed vertical and lateral loads. Because the purpose of a retaining wall is to hold back a mass of soil or other material, the design of the wall is affected by the earth's configuration—whether the earthen surface behind the wall is horizontal or inclined. Design is also affected by any additional loading, as from a vehicle or equipment passing near the top of the wall, which causes horizontal thrust on the retaining wall.

There are three basic types of concrete masonry retaining wall: gravity, cantilever, and counterfort or buttressed walls.

A *gravity* wall depends on its own mass for stability. Basically, it is massive masonry laid so that little or no tension stress occurs in the wall under

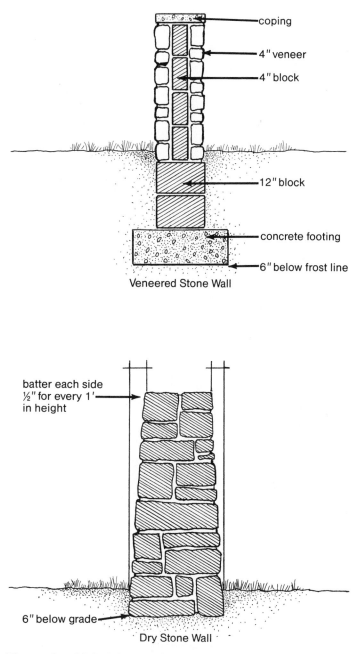

Veneered Stone Wall

Dry Stone Wall

Illustration 13-9. A freestanding stone wall can be constructed using stone veneer on a concrete block wall or by using stone constructed with or without mortar. In mortarless construction, no footing is required, but in mortar construction, a footing below the frost line is necessary.

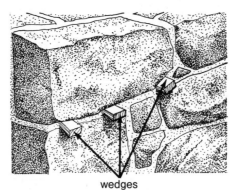

wedges

Illustration 13-10. Because stones are very heavy, it is necessary to use wooden wedges to support them while the mortar stiffens. Once the mortar has set, remove the wedges and tuckpoint the holes.

loading, and its cross section is usually a trapezoidal shape (see illustration 13-11). Because a retaining wall of the gravity type ordinarily has a base thickness equal to one-half to three-fourths the wall height, it is usually more economical to use a cantilever design for a large retaining wall (one more than a few feet in height).

A *cantilever* wall usually has a cross section shaped like an inverted T (see illustration 13-12), with the stem located toward the rear of the footing if soil-bearing stresses are critical and toward the front or toe if sliding is

critical. An L-shaped cantilever wall is used along a property line or in other situations where it is impossible to provide a toe; for such a wall, bearing pressure is usually high. (The dimensions shown by the letters H, a, b, and t in illustration 13-12 refer to the data provided in table 13-7.)

Counterfort or *buttressed* retaining walls are similar to cantilever walls except that they span horizontally between vertical supports. Supports at the back of the wall are known as counterforts (see illustration 13-13), and those supports exposed at the front are called buttresses (see illustration 13-14).

A small degree of forward or outward tilt under service conditions is difficult to avoid with any type of retaining wall. It is therefore good practice to batter (slope) the front face of the wall slightly to offset this tilt and avoid the illusion of instability. A batter on the order of $1/2$ inch per foot of wall height is commonly used for masonry construction.

The selection of a particular type of retaining wall for cost and efficiency depends on the wall size, loads, soil conditions, and site location. The cantilever type of wall has a slightly lower toe pressure than the gravity type and thus may be desirable where soil-bearing capacity is low. However, the gravity wall has greater resistance to sliding because of its greater weight.

Construction

The construction of the footing in a retaining wall is important, and should be placed on firm, undisturbed soil. In areas where freezing temperatures are expected, the base of the footing is placed below the frost line. Where soil under the footing consists of soft or silty clay, 4 to 6 inches of consolidated granular fill can be placed under the footing slab to assure firm support and to increase the frictional resistance

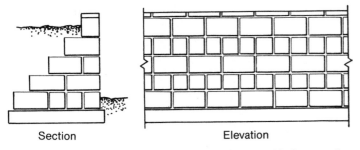

Section   Elevation

Illustration 13-11. A gravity retaining wall gains added strength from the weight of the earth on the structure.

between the footing and the ground. This friction determines resistance to horizontal sliding of the wall.

Often a lug or key under the footing is provided to assist in resisting the tendency to slide (see illustration 13-15). The same effect is achieved by requiring that the footing be well below the excavated surface in undisturbed soil, particularly if the wall is higher than 7 feet above the footing.

The top of the concrete footing in the area under the masonry is roughened while the concrete is still fresh. Otherwise, a 1-inch-deep, 4-inch-wide keyway is provided to improve shear bond at the joint between the wall and the footing.

The materials and procedures previously recommended for reinforced grouted masonry walls (see Chapters 10 and 11) should be followed in the construction of retaining walls. It is necessary to provide some horizontal steel reinforcement to distribute stresses that occur when the wall expands or contracts. The amount of horizontal reinforcement needed is, to a large extent, dependent on climatic conditions. For moderate conditions and 8-inch walls, bond beams with two #4 bars should be placed in the top course and in intermediate courses at 16 inches on center. For 12-inch walls, the top bond beam should contain two #5 bars and the intermediate bond beams should have two #4 bars. If desired, horizontal joint reinforcement may be placed in each joint (8 inches on center) and the bond beams omitted.

Drainage

Provisions must be made to prevent accumulation of water behind a retaining wall. Water accumulation causes increased soil pressure, seepage, and, in areas subject to frost action, expansive forces of considerable magnitude near the top of the wall.

Table 13-7. **Reinforcement Requirements for Cantilevered Walls***

Block Width	H*	a*	b*	t*	Dowel and Vertical Reinforcement	Top Footing Reinforcement
8"	3'4"	12"	2'8"	9"	#3 @ 32" o.c.	#3 @ 27" o.c.
	4'0"	12"	3'0"	9"	#4 @ 32" o.c.	#3 @ 27" o.c.
	4'8"	12"	3'3"	10"	#5 @ 32" o.c.	#3 @ 27" o.c.
12"	3'4"	12"	2'8"	9"	#3 @ 32" o.c.	#3 @ 27" o.c.
	4'0"	12"	3'0"	9"	#3 @ 32" o.c.	#3 @ 27" o.c.
	4'8"	12"	3'3"	10"	#4 @ 32" o.c.	#3 @ 27" o.c.
	5'4"	14"	3'8"	10"	#4 @ 24" o.c.	#3 @ 25" o.c.

NOTES:

Reinforcing bars should have standard deformations.

Alternate vertical reinforcing bars may be terminated at the midheight of the wall. Every third bar may be terminated at the upper third-point of the wall height.

The wall should have horizontal joint reinforcement at every course or else a horizontal bond beam with two #4 bars every 16 inches.

*The letters H, a, b, and t refer to wall dimensions shown in illustration 13-12.

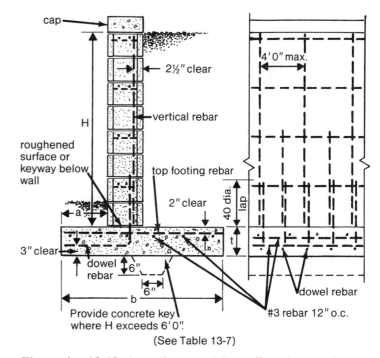

Illustration 13-12. A cantilever retaining wall requires continuous reinforcement between the footing and the wall.

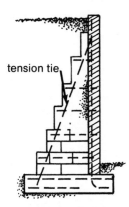

Illustration 13-13. A counterfort retaining wall gains added strength by means of reinforced extensions protruding into the soil bank.

As shown in illustration 13-15, 4-inch-diameter weep holes spaced 5 to 10 feet along the base of the wall should provide sufficient drainage of permeable backfill soils. An alternative type weep hole can be created by leaving the mortar out of several head joints in the first or second course, and placing 1 cubic foot of gravel or crushed stone behind each of these joints.

Where unusual conditions such as heavy, prolonged rains will be encountered, seepage through weep holes may cause the ground in front of the wall and under the toe of the footing to become saturated and lose some of its bearing capacity. This undesirable condition can be avoided by waterproofing the back of the wall, using gravel as backfill immediately next to the wall, and installing a continuous horizontal run of drain tile along the base of the wall. If the drain is surrounded by crushed stone or gravel, extended the full length of the back of the wall and provided with outlets beyond the ends of the wall, no weep holes are needed in the wall itself.

The top of a concrete masonry retaining wall should be capped or otherwise protected to prevent entry of water into unfilled hollow cores and spaces. Climate and type of construction will determine the need for waterproofing the back face of the wall. Because saturated mortar may not be durable in areas subject to frequent freezing and thawing, waterproofing is recommended when backfill material is relatively impermeable; it is also recommended to reduce unsightly efflorescence or leaching on the wall.

A long retaining wall should be broken into panels by means of vertical control joints. The joints should resist shear and other lateral forces in order to maintain alignment of adjacent wall sections while permitting longitudinal movement (see illustration 13-16).

Backfilling should not be permitted until at least 7 days after grouting. It is good practice to build up the backfill material all along the wall at a rate as nearly uniform as practicable. If heavy equipment is used in backfilling a wall designed to resist only earth

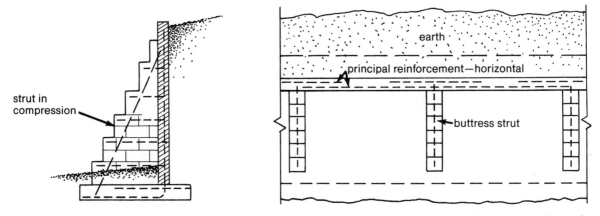

Illustration 13-14. A buttressed retaining wall is like a counterfort structure except that the struts are located on the front side of the wall.

pressure, such equipment should not approach the back of the wall closer than a distance equal to the height of the wall. Care should also be taken to avoid large impacting forces on the wall such as those caused by a large mass of moving earth or large stones.

Where the finished grade at the back of a retaining wall is level or nearly so, a fence or railing on top of the wall may be needed for safety. One way to accomplish this is to build the masonry wall itself higher by using screen block units.

Stone Retaining Walls

Stone retaining walls are simple to build but require more work in trying to fit the irregular pieces into an aesthetic overall architectural element. There are two ways to construct a stone retaining wall—dry or wet. Dry construction means that no mortar is used in the construction of the wall; it is relatively simple to do but also challenging.

A dry retaining wall relies on gravity and friction to hold it together, there-

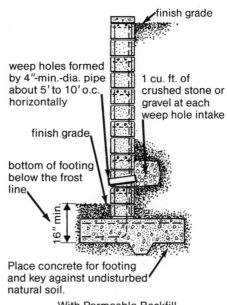

finish grade

weep holes formed by 4"-min.-dia. pipe about 5' to 10' o.c. horizontally

1 cu. ft. of crushed stone or gravel at each weep hole intake

finish grade

bottom of footing below the frost line

16" min.

Place concrete for footing and key against undisturbed natural soil.

With Permeable Backfill

Illustration 13-15. Drainage for a retaining wall is critical and provision must be made for it by using appropriate waterproofing, backfill, and water outlets.

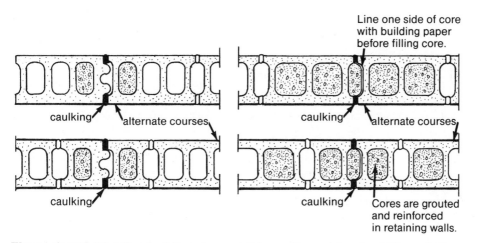

Line one side of core with building paper before filling core.

caulking   alternate courses

caulking   alternate courses

caulking

caulking   Cores are grouted and reinforced in retaining walls.

Illustration 13-16. When building a long retaining wall, provision should be made for control joints that will resist shear.

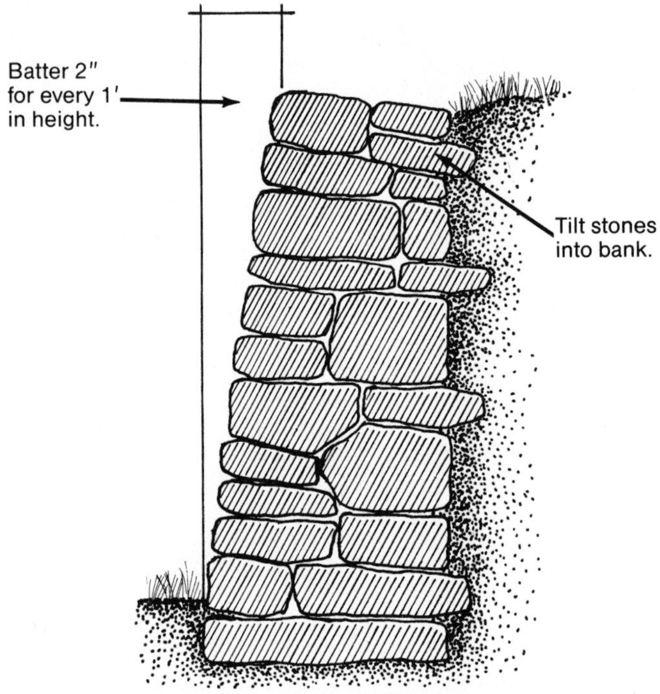

Batter 2"
for every 1'
in height.

Tilt stones
into bank.

Illustration 13-17. Stone retaining walls, like freestanding walls, can be built using either mortar or mortarless construction. When building a mortarless wall, allow some long stones to extend into the bank.

stones providing a stable base. And be careful when laying up the stones that you have as many broken lines as possible—continuous lines are not as aesthetic nor are they as structurally sound. Backfill the earth as you progress and use larger stones protruding into the backfill to help secure the wall.

Wet retaining walls require mortar between the stones and may be constructed in one of two ways—either as a solid stone wall or as a veneer to concrete block or a poured concrete retaining wall (see illustration 13-18). For a solid, wet stone retaining wall, the first course must be placed 6 inches below the frost line and a 6-inch-diameter drain must be provided behind the wall and 6 inches below grade as in a dry wall. Because of the mortar, weep holes must be provided in the wall to allow water to pass through.

Construction for a mortared stone retaining wall is done in the same manner as a masonry retaining wall. A mortared stone retaining wall like a dry retaining wall must be battered. Because mortar instead of gravity is holding the wall together, the batter need only be 1 inch for every foot of height. If you are using stone as a veneer, construct a masonry wall as described earlier and embed metal tabs in the mortar joints. Lay a 4-inch stone veneer facing beginning on top of the footing and extend it to the top of the masonry block wall. Alternately, the veneer can be supported on concrete block below grade (see illustration 13-18). A cap of precast units or stone completes the structure.

Sidewalks, Patios, and Driveways

One of the easiest masonry projects to undertake is the paving of

fore any wall over 2 feet high requires a batter of 2 inches for every foot of height (see illustration 13-17). No footing is required for a dry stone retaining wall because the stone wall will rise and fall with the freezing and thawing of the earth, but you need to begin the wall 6 inches below grade. If your plans call for a 3-foot-high wall, the base should be at least 18-inches wide so that you will end up with a 1-foot dimension on the top of the wall.

To construct the wall, drive batter boards into the ground about every 4 to 6 feet. Lay stones the way they are found on the ground—flat. Turning stones on end creates an unstable wall. The first course should consist of larger

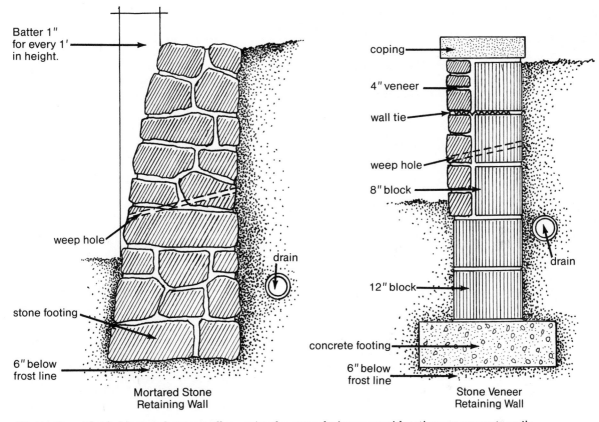

Batter 1"
for every 1'
in height.

weep hole

stone footing

6" below
frost line

**Mortared Stone
Retaining Wall**

coping

4" veneer

wall tie

weep hole

8" block

drain

12" block

concrete footing

6" below
frost line

drain

**Stone Veneer
Retaining Wall**

Illustration 13-18. Mortared stone walls require the same drainage considerations as concrete walls.

sidewalks, patios, and drives with masonry units or stone. The variety of materials available and the combinations possible are almost unlimited.

Once you decide to undertake a paving project for your home, you will need to decide which kind of material best suits your design and your budget. Availability of material varies from area to area and from one supplier to the next. So visit several suppliers and compare their products and unit prices.

Precast concrete paving slabs and interlocking pavers are machine-made or precast by conventional methods. Slabs are commonly available in square and other rectangular shapes measur-

ing from 12 to 36 inches, with thickness of 2, 2½, or 3 inches. The 2-inch thickness is suitable for residential walks and patios, while the 2½-or 3-inch thickness should be used for driveways. Other sizes and shapes are also manufactured: round, triangular, diamond, hexagonal and Spanish tile.

Interlocking concrete pavers are growing in popularity and use at a very rapid rate all over the world. They are made in a number of shapes and colors and are ideal for do-it-yourself applications. Concrete pavers interlock without the use of mortar. This allows them to be "unzipped" for work on or beneath the subsurface, then "zipped"

Interlocking concrete paving units are durable, easy to replace or remove, and add a touch of elegance to a home.

back using the same material. Typical paver thicknesses are $2^3/8$ inches (60 mm) and $3^1/8$ inches (80 mm). The thinner units are for pedestrian areas and residential driveways, the $3^1/8$-inch pavers are for streets and other vehicular applications.

Architects and designers are now using interlocking concrete pavers in many public and private areas. These pavers have been around for years in Europe, are easy to use, and come in a wide variety of color and shape. The advantage in using them is that no one unit can tip out of alignment without taking several adjoining units with it. This helps the surface stay intact even under heavy loads.

One special type of paving unit is the turf-retaining block (see illustration 13-19). This unit is designed to maximize green space by allowing grass to grow in paved areas.

The dense and solid brick used in sidewalks, patios, and driveways are

also called *pavers*. Those best for all-purpose use and varied weather conditions are rated as Class SX, Type 1. Used brick are not recommended for paving because they may be porous and susceptible to deterioration under freeze-thaw conditions.

Brick pavers are generally available in sizes measuring from a nominal 4×8 inches to an actual 4×8 inches. Also available are 6- and 8-inch hexagonal units. These pavers are available in thicknesses of $1^5/8$ to $2^1/4$ inches with the $2^1/4$ size being the most widely used.

Stone is the oldest and also the most natural and rugged of all paving materials. Popular types chosen for paving are limestone, sandstone, granite, and slate. The large, flat stones most often used for paving are called *flagstones*, whereas, the small, roughly rounded stones are called *cobblestones*. Because of shipping costs and limited availability, in some areas of the country stone is quite expensive. Still, for many homeowners, its natural beauty and strength outweigh the cost.

Design Considerations

Designing a walk, patio, or driveway requires the same considerations as discussed in Chapter 3 for cast-in-place concrete. Slopes, widths, and drainage must all be taken into account with modular paving. The only difference is that you need to determine ahead of time the type of paving material you want to use, then design your project to gain the maximum usage of those paving units. Adjusting a plan 2 or 3 inches in one direction may save you hours of extra cutting and fitting.

Illustrations 13-20, 13-21, and 13-22 show just some of the patterns possible with concrete pavers, brick, and stone respectively.

Construction Techniques

There are two ways to install paving units for sidewalks, patios, and driveways—mortared or unmortared. Either method can be used to create a durable, long-lasting surface. But, regardless of the method used, the final job will be only as stable as the subgrade underneath the sand or mortar. Subgrade preparation should be done as carefully for paving work as for cast-in-place work.

Pavement construction can begin in spring as soon as the frost is gone and ground conditions permit. In the fall, installation can continue until frozen ground prevents proper compaction of the subgrade material.

Preparing the Subgrade The most important thing to consider when

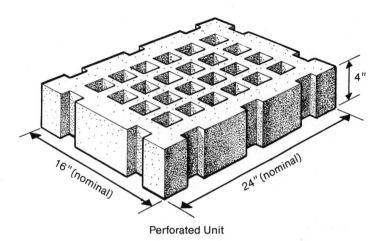

Perforated Unit

Illustration 13-19. Orginally used for erosion control, perforated units allow grass to grow in openings thereby permitting your pavement to double as green space.

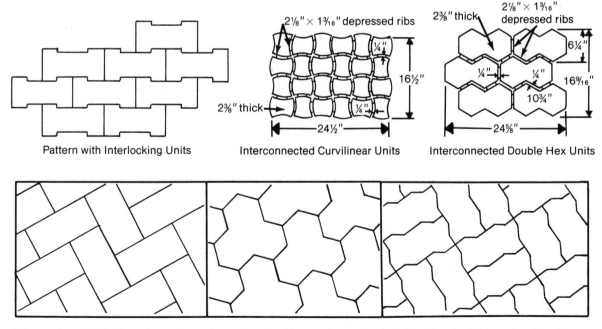

Pattern with Interlocking Units   Interconnected Curvilinear Units   Interconnected Double Hex Units

Illustration 13-20. There is a wide variety of interlocking paving units available. Check with your local building supplier.

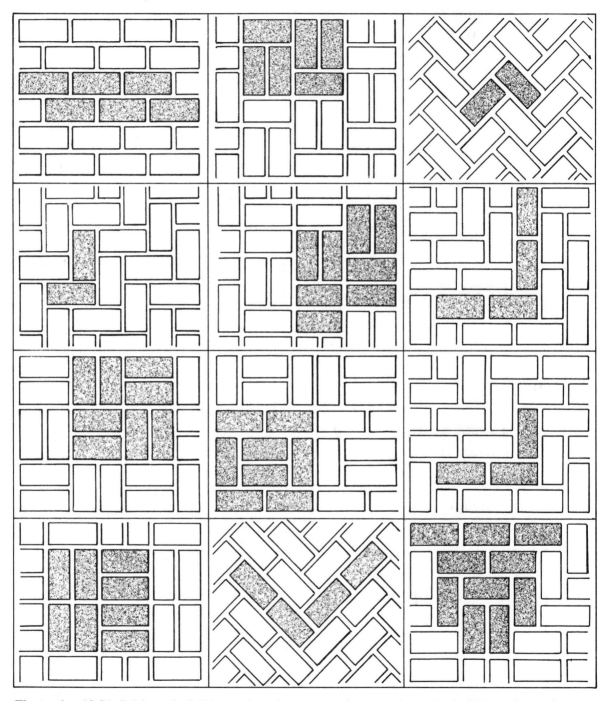

Illustration 13-21. Brick may be laid in a variety of patterns, and patterns intermixed within paving sections.

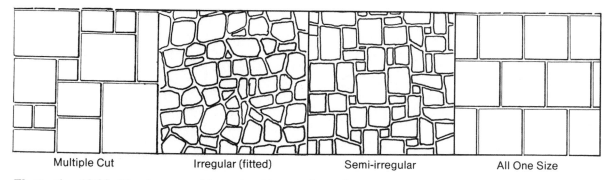

Multiple Cut   Irregular (fitted)   Semi-irregular   All One Size

Illustration 13-22. Flagstones provide a natural beauty that enhances most any home, and they are available in both cut and uncut form.

preparing the subgrade is drainage. Water must be carried away from the paved area and this is done by grading. To grade an area you should start by setting stakes at intervals of 5 to 10 feet. On one stake, mark the height of the finished pavement (which should be at or above the existing grade). Then mark the other stakes, allowing for a pitch of 1 inch of slope for every 10 feet of length.

Next, you need to excavate the area, removing the soil to a depth equal to the thickness of the paving unit plus $1\frac{1}{2}$ to 2 inches for the sand base. Be careful not to remove too much soil, because undisturbed soil makes the best subgrade and is superior to soil that has been dug out and replaced. If you are installing a stone walk or patio using large flagstones, you need only remove the topsoil, because stones may rest directly on the ground if the soil is already well drained.

Edgings   Retaining edges are required to prevent horizontal movement of mortarless bricks and other paving units. There are several types of edgings you can construct depending on your design preferences (see illustration 13-23).

Cast-in-place concrete edges are installed in the same manner as footings. They should extend 8 to 12 inches below the surface on a 2- to 4-inch bed of sand or directly on well-drained soil. Brick edging is set in soil, and is made by standing the brick on end with faces toward the pavement or laying them on edge against the pavement. Wooden edgings can be made of 2×6 redwood, cedar, or pressure-treated wood held in place by stakes placed approximately every 4 feet. Stakes should be driven down 2 inches below the top edge of the frame and beveled sharply down from that edge so they can be covered with soil.

Put edgings in place before creating a sand base and laying pavers unless you cast concrete curbs after installing pavers (see illustration 13-23). Use string and stakes to maintain straight lines and correct dimensions as you work.

Base Preparation   Base preparation varies depending on the type of subgrade and the method of construction. Mortarless construction requires the simplest preparation: $1\frac{1}{2}$ to 2 inches of sand is all that is needed to create a base for paving units. If the base is in

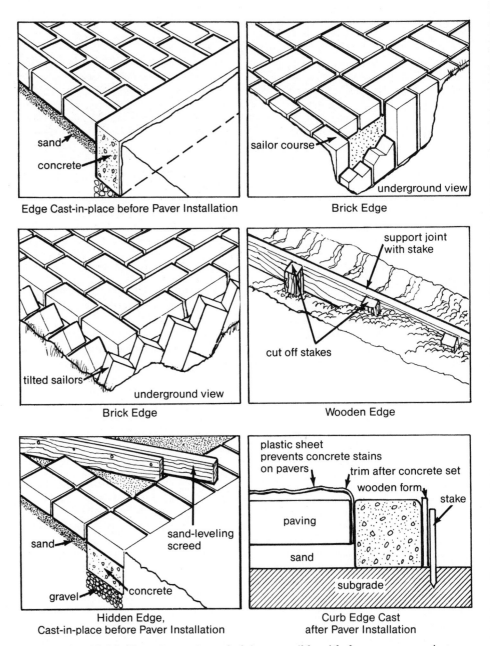

Edge Cast-in-place before Paver Installation

Brick Edge

Brick Edge

Wooden Edge

Hidden Edge,
Cast-in-place before Paver Installation

Curb Edge Cast
after Paver Installation

Illustration 13-23. There is a variety of edgings possible with dry masonry paving construction.

an area of poor drainage, excavate the subgrade an additional 2 to 4 inches and fill the entire base with pea gravel or larger stone gravel. Pea gravel can be easily screeded, but a larger gravel is less expensive.

The sand used for the base should be of concrete sand quality. If possible, when the sand is delivered have it placed in small piles on the subgrade to minimize further handling. When ordering sand for bedding and joint filling, allow about 1 cubic yard for every 150 square feet of paving area.

After the sand is delivered and moved into place, level it with a screed. Once the sand has been screeded, stay off of it. If the prepared bed does get walked on, rework and screed the area again.

For mortared construction, paving units must be placed on a new or existing concrete slab. If you need to construct a new slab, follow the procedures set forth in Chapter 3.

Begin the base for mortared construction by laying a bed of mortar made of a 1:4 cement-sand mix. After placing the mortar, screed it in the same manner described for a sand base. Be sure to mix only as much mortar as you will be able to use in an hour and screed only a 10 square foot area at a time.

Placing Units   Placing the first few paving units requires special care because they determine the line and grade for placing the rest. As you proceed with each row, use string lines to serve as a guide in order to keep the pavement straight (see illustration 13-24). Begin laying the units from a corner, and place them from only one direction. If you have a helper, you should both be working together at one place and in the same direction.

When working with interlocking

Concrete pavers are available in many sizes, shapes, and colors.

concrete pavers, check to see how they fit together. They may have to be laid in a certain sequence to stay even. Also, note that no mortar should be used with the interlocking type of pavers. They are designed to fit hand tight without mortar.

If you use mortarless construction, butt concrete and brick pavers against each other hand tight, and stop laying pavers approximately 20 inches short of the screeded sand edge. Irregular paving units, such as flagstone, cannot be butted against each other; however, avoid large gaps by selecting and fitting stones that are a close match.

When laying pavers that will be mortared, use a wooden spacer to maintain a 1/2-inch gap between all units. Before inserting the mortar, check all units with a level. Even them up by tapping with the trowel handle (see illustration 13-24). To avoid uneven compaction of the base, use a piece of

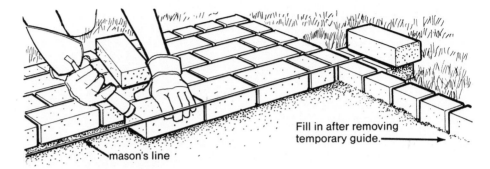

Fill in after removing
temporary guide. ➤

mason's line

Illustration 13-24. Always use a string to keep bricks in a straight line and fit bricks snugly by tapping them with a trowel handle.

Illustration 13-25. In dry mortar work, use a bristle brush to sweep the mixture into the joints. Use a piece of plywood to support your weight so it will not displace any of the units.

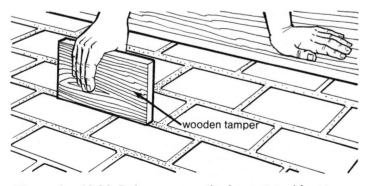

wooden tamper

Illustration 13-26. Before you spray the dry mortar with water, tamp the mixture into the joints with a ½-inch piece of wood.

¾-inch plywood as a kneeling board to distribute your weight.

Finishing   There are two ways to mortar paving units once they are laid. The first method is a dry method where a 1:4 mix of cement and dry sand is dry mixed, placed on the pavement, and brushed into the joints. When using this method, be sure to use a kneeling board and carefully sweep all mortar from the upper surface of the bricks (see illustration 13-25). Then, take a ½-inch-thick piece of wood and tamp all the mortar in the joints (see illustration 13-26). Add more dry mix if needed to completely fill the joints.

Again, sweep any mortar on the units into the joints. Finally, wet the surface using a very fine water spray. But, be careful not to splash any mortar out of the joints and don't allow any water to pool. Keep the mortar damp over the next 2 to 3 hours; then, when the mortar is firm, tool the joints with a jointer as you would when finishing a masonry wall. After a few more hours, scrub off any mortar on the units with burlap.

The alternative method is to mix mortar in the standard fashion and trowel the mix into the joints. When using this method, work carefully to minimize spilling and clean all excess mortar off the face of the paving units. When the mortar begins to harden, tool the joints.

Finishing is a bit different in mortarless construction. Once all the units are laid, spread dry sand over the entire surface. If the sand is not dry, allow it to sit for a while so it can dry. Then, broom the sand into the cracks. If you wish, rent a vibrator and go over the entire paved area packing the masonry units evenly into the sand bed. Only use a vibrator after sand has been swept into the joints; and, after vibrating, fill the cracks a second time.

Cleaning Masonry

If you choose not to apply a finish to a masonry construction, you'll need to clean off the mortar smears resulting from your work. As noted in the previous chapters, lumps of mortar can be knocked off after they've almost dried, but mortar smears require a cleaning solution of muriatic acid and water. When mixing the acid with the water, always pour the acid slowly into the water— *never* pour the water into the acid. Acid also requires safe handling so you must use it in a well-ventilated area and wear rubber gloves, eye protection, a long-sleeved shirt and full-length trousers.

To use the mixture, apply the acid to a small area, let it stand for 3 or 4 minutes, then flush thoroughly with clean water. Because acid may discolor a masonry surface, test it on an inconspicuous area first. For concrete, concrete block, and dark-colored brick, use a 1:9 acid-water solution. This solution may discolor light-colored masonry, so use a 1:14 or 1:19 solution. Never use acid on colored mortars or concrete because it may leach out the color. Never use acid on stone. For removing other stains from masonry construction, refer to Chapter 7.

Glossary

Absorption—The process of moisture filling permeable pores in a porous solid material.

Accelerator—An admixture that speeds the rate of hydration of cement, shortens the normal time of setting, or increases the rate of hardening, strength development, or both, of concrete, mortar, or grout.

Acrylic resin—A synthetic resin used in concrete construction as a surface sealer or a bonding compound.

Addition—A substance that is interground or blended in limited amounts into a hydraulic cement during manufacture—not on the job—either as a processing addition to aid in manufacturing and handling the cement or as a functional addition to modify the use properties of the cement. (Improperly called additive.)

Admixture—A material, other than water, aggregate, or cement, used as an ingredient in concrete, mortar, or plaster and added to the mix either immediately before or during the mixing.

Adobe—Unfired clay bricks that are dried in the sun. A common building material in the Southwest and closely simulated by a type of concrete masonry unit called slump block.

Aggregate—Materials such as sand, gravel, and crushed stone used with cement to make concrete or mortar.

Agitating truck—Another term for a concrete truck. It is usually a truck with a rotating drum that continuously agitates fresh concrete during transport from the ready-mix plant to the work site.

Air content—The volume of air voids in cement paste, concrete, or mortar. Entrained air adds to the durability of hardened concrete and the workability of fresh concrete and mortar mixes.

Air-entraining agent—An admixture for concrete or mortar that causes the formation of air bubbles in the mix in order to improve workability and frost resistance.

Air entrainment—The introduction of air in the form of minute bubbles (usually smaller than 1 millimeter in diameter) into concrete or mortar during the mixing in order to improve the flow and workability of fresh mixes as well as the durability of the hardened material.

Air void—An entrapped or entrained air pocket in concrete or mortar. Entrapped voids usually are larger than 1 millimeter in diameter; entrained air voids are smaller. Entrapped air voids should be removed with vibration, power screeding, or rodding.

Anchor bolt—A headed or threaded metal bolt or stud that is either cast-in-place, grouted in place, or cemented into a drilled hole, and used to attach steel or wooden structural members to concrete.

Architectural concrete—Concrete that is permanently exposed to view and requiring special care in selection, placing, and finishing in order to get the desired architectural effect.

Ashlar—Usually a term referring to squared stones but also used as a name of a pattern in masonry construction.

Backfill—Earth, rubble, etc., used to correct overexcavation or to replace earth in a trench or around a foundation wall.

Back plastering—Applying a backup coat (or coats) of plaster to the back side of a solid plaster partition after the wall mortar or plaster on the opposite side has hardened.

Bag—A measure of cement equal to 94 pounds in the United States.

Base coat—Plaster coat or coats applied before the final coat.

Batch—The quantity of concrete or mortar that is mixed at one time.

Batching—Weighing or volumetrically measuring and introducing into the mixer the ingredients for a batch of concrete or mortar.

Bed joint—A horizontal joint in a masonry wall.

Bleeding—The flow of mixing water within, or its emergence on the surface of newly placed concrete or mortar; caused by the settlement of the solid materials within the mass.

Block—A concrete masonry unit, usually containing hollow cores.

Bond—Adhesion of concrete, mortar, or plaster to other surfaces against which they are applied. Some examples are adhesion of cement paste to aggregate, adherence between plaster coats or between plaster and a substrate, and adhesion of concrete to reinforcing steel.

Bonding agent—A substance applied as a coating to a suitable substrate to create a bond between the substrate and a succeeding layer of concrete.

Brown coat—The second coat of three-coat plastering.

Buck—Framing used around an opening in a wall where a door or window will be placed.

Bug holes—Small holes in concrete caused by air bubbles trapped in the surface of formed concrete during placement and compaction.

Bulking—The increase in the volume of a quantity of sand in a moist condition compared to the volume of the same quantity of sand in a dry state.

Bull float—A tool with a large, flat, rectangular piece of aluminum, wood, or magnesium measuring 8-inches wide × 42- to 60-inches long with a 4- to 16-foot handle used to smooth concrete slabs.

Bush-hammer—A hammer with a serrated face used to roughen a concrete surface.

Buttering–The process of spreading mortar on a masonry unit with a trowel.

Cast-in-place–Concrete or mortar that is deposited in the place where it will harden as part of a structure (opposite of precast).

Cement, masonry–A cement for use in plaster and in mortars, containing one or more of the following materials: portland cement, blended cement, natural cement, slag cement or hydraulic lime; and in addition usually containing one or more materials such as hydrated lime, limestone, chalk, as prepared for this purpose.

Cement paint–A paint made mostly of white portland cement, water, and lime.

Cement paste–A constituent of concrete consisting of cement and water.

Checking–Development of shallow cracks at closely spaced but irregular intervals in concrete, plaster, or mortar surfaces; also known as crazing or craze cracks.

Chemical bond–The bond between materials resulting from cohesion and adhesion developed by chemical reaction.

Clinker–A partially fused product of a kiln, which is ground to make cement.

Closure–A whole or partial masonry unit that is used to complete a course in a masonry wall.

Cobblestone–A rock fragment that is usually rounded or semirounded and averaging between 3 to 12 inches in diameter.

Cohesion–The ability of a material to cling to itself.

Compaction–The process of reducing the volume of freshly placed concrete or mortar by vibration, tamping, or rodding in order to ensure complete embedment of all reinforcement and to eliminate entrapped voids.

Concrete, green–Concrete that has been placed and has set but is not completely hardened.

Consistency–The relative mobility or ability of freshly mixed concrete or mortar to flow, measured by slump in concrete.

Construction joint–The surface where two successive concrete placements meet.

Control joint–A weakened plane in concrete, plaster, or masonry construction, created by a formed, sawed, or tooled groove. Control joints, also called contraction joints, avoid the development of high stresses and regulate the location of possible cracking due to expansion, contraction, settlement, and initial drying shrinkage.

Coping–Materials or units used to form a cap or finish on top of a wall or chimney.

Curing–Keeping freshly placed concrete, plaster, or mortar moist and at a favorable temperature for a suitable length of time to assure satisfactory hydration.

Curing compound–A membrane-forming liquid applied as a coating to surface of newly placed concrete in order to retard the loss of water.

Curling–The distortion or warping of an essentially flat surface into a slightly curved shape, owing to several factors, but primarily due to moisture differences between the top and bottom of a concrete slab.

Damp proofing–Treatment of concrete, masonry, or plaster to retard the passage or absorption of water, or water vapor, either by application of a suitable coating to exposed surfaces or by use of a suitable film, such as polyethylene under slabs on grade.

Darby—A hand-manipulated straightedge, usually 3 to 8 feet long, used in the early stage leveling of concrete or plaster surfaces.

Deformed bar—A reinforcing bar with ridges that serve to lock into surrounding concrete.

Deicer—A chemical such as sodium or calcium chloride used to melt snow and ice on pavements.

Devil's float—A wooden float with two nails protruding from the toe used to roughen the surface of a brown coat in plastering and stuccoing.

Diamond mesh—Sheets of metal that are slit and pulled out to form diamond-shaped openings; used as metal reinforcement for plaster. Also known as expanded metal lath.

Divider strips—Nonferrous or plastic strips used in terrazzo work and embedded to depths of 5/8 to 1 1/4 inch. Also may be 1 × 4 or 2 × 4 redwood, cypress, or cedar left in concrete slabs permanently as a decorative or control joint.

Double-headed nail—A nail with two heads used in formwork. The nail is driven into the first head and removed by the protruding second head.

Dowel—Steel pin extending into adjoining portions of concrete to prevent shifting of the concrete.

Dry mix—Prepackaged concrete, mortar, or plaster mixtures usually sold in bags and containing all ingredients except water.

Dry-shake—A dry mixture of cement and fine aggregate (usually colored) that is spread on a concrete surface after bleed water has disappeared following strikeoff and worked into the surface during floating.

Durability—The ability of portland cement concrete and mortar to resist weathering action, chemical attack, abrasion, etc.

Dusting—The development of powdered material on the surface of concrete that has hardened.

Efflorescence—A deposit of salts, usually white, formed on a surface; the substance emerging in solution from within the concrete, masonry, or plaster and deposited by evaporation.

Falsework—Temporary structures such as shoring, formwork for beams and slabs, etc., built to support work in progress.

Fin—A narrow projection on a concrete surface caused by mortar flowing between the cracks in the forms.

Finish coat—The last coat of plaster, the decorative surface, usually is colored and frequently textured.

Finishing—The leveling, smoothing, compacting, and treatment of the surface of concrete, plaster, etc. to attain the final desired texture.

Float—A rectangular hand tool, usually of wood, aluminum, or magnesium, used to impart a relatively even but still open texture to a concrete surface. Other types—cork, sponge rubber, etc.—are used in plastering and grout work.

Fog curing—The application of a fine mist of water used during the curing of concrete, mortar, or stucco.

Form—A temporary structure or mold used to contain fresh concrete while it hardens.

Form oil—Oil applied to the interior of the form to promote the release of the form from the concrete.

Frog—A depression in the bed surface of a brick or other masonry unit.

Gradation—The size distribution of aggregate particles, determined by separation with standard screen sieves.

Groove joint—A control joint made by forming a groove in the concrete surface before hardening to control crack location.

Grout—A mixture of cementitious material and water with or without aggregate, that is usually proportioned to produce a pourable consistency that will not segregate.

Hardener—A chemical applied to a concrete floor to reduce wear and dusting. Dry-shakes used to increase wear resistance are also sometimes called hardeners.

Harsh mixture—A mixture that lacks desired workability and consistency due to a deficiency of cement paste, aggregate fines, or water.

Hawk—A tool to hold and carry plaster or mortar; generally a flat piece of metal approximately 10- to 14-inches square, with a wooden handle fixed to the center of the underside.

Header—A masonry unit placed perpendicular (across the thickness) to a wall to bond two or more wythes of a wall together by overlapping.

Metal lath—*See* Diamond mesh.

Mortar—A mixture of cement paste and fine aggregate.

Pointing—The process of repairing mortar joints by removing old mortar and adding new; also called tuckpointing.

Raking—The removal of slightly hardened mortar from a joint in a masonry wall to accent the line of the joint.

Rebar—Reinforcing steel bars or rods.

Retardation—Slowing down the rate of hardening or setting, usually in hot weather, to gain an increase in the time required to reach initial and final set or to develop early strength of concrete, mortar or plaster.

Retarder—An admixture that delays the setting of cement paste and hence of concrete, mortar, or plaster.

Retempering—Adding water and re-mixing concrete, mortar, or plaster that has started to stiffen and become harsh.

Rowlock—A header unit that is set on its edge.

Sailor—A brick or block that is standing on end with broad face showing.

Scarifier—A tool with flexible steel tines used to scratch or rake the unset surface of a first coat of plaster. Also refers to a tool used to roughen the surface of hardened concrete.

Scoring—Grooving, usually horizontal, of the scratch coat of stucco to provide a mechanical key with the brown coat.

Soldier—A brick or block that is standing on end with narrow face showing.

Stretcher—A masonry unit that is laid parallel with the wall.

Striking—The cutting away of mortar with the trowel; also the tooling of mortar joints.

Trowel—A flat, broad-bladed steel hand tool used to finish concrete, or to apply, shape, and finish plaster.

Tuckpointing. See Pointing

Vapor barrier—A material used to retard the flow of moisture into walls or through concrete floor slabs that are cast on the ground.

Vibrator—A machine used to eliminate trapped air bubbles and consolidate freshly placed concrete.

Wale—A long horizontal member on formwork used to hold studs in place.

Wall tie—A metal reinforcing strip or wire used to attach veneers or hold wythes together.

Warehouse set—The partial hardening of bagged cement or prepackaged mixes caused by the absorption of atmospheric moisture during improper or lengthy storage. Also called warehouse pack.

Water-reducing agent—A material that increases slump and workability of freshly mixed concrete or mortar without increasing the amount of water.

Workability—The property of freshly mixed concrete, mortar, or plaster that determines its working characteristics and the ease with which it can be mixed, placed, and finished.

Wythe—A continuous vertical section of a wall that is one masonry unit wide.

For Further Reference

More complete information on many of the topics discussed in this book is found in the following books published by the Portland Cement Association:

- *Cement Mason's Guide*
- *Concrete Masonry Handbook*
- *Design and Control of Concrete Mixtures*
- *Portland Cement Plaster (Stucco) Manual*

For a complete, up-to-date catalog of these and other books, bulletins, microcomputer programs, slide sets, films, and video cassettes, send your request to the following address:

Portland Cement Association
Order Processing
5420 Old Orchard Road
Skokie, IL 60077–9973

Index

Page references in *italics* indicate illustrations or photographs.

221